T0401670

THOMAS D. GRANT

NUCLEAR ARMS CONTROL IN PERIL

Why the Nuclear Non-Proliferation Treaty Matters and How to Save It

First published in Great Britain in 2025 by

Bristol University Press
University of Bristol
1–9 Old Park Hill
Bristol
BS2 8BB
UK
t: +44 (0)117 374 6645
e: bup-info@bristol.ac.uk

Details of international sales and distribution partners are available at bristoluniversitypress.co.uk

British Library Cataloguing in Publication Data
A catalogue record for this book is available from the British Library

ISBN 978-1-5292-4779-4 hardcover
ISBN 978-1-5292-4780-0 ePub
ISBN 978-1-5292-4781-7 ePdf

Cover design: Blu Inc
Front cover image: stockcake.com
Bristol University Press uses environmentally responsible print partners.
Printed in Great Britain by CPI Group (UK) Ltd, Croydon, CR0 4YY

Contents

List of Abbreviations

CD	Conference on Disarmament
CEND	Creating an Environment for Nuclear Disarmament
CTBT	Comprehensive Nuclear-Test-Ban Treaty
FMCT	Fissile Material Cut-off Treaty
GAR	General Assembly Resolution (UN)
HoC	House of Commons (UK)
HoL	House of Lords (UK)
IAEA	International Atomic Energy Agency
ICJ	International Court of Justice
ILC	International Law Commission (UN)
INF Treaty	Intermediate-Range Nuclear Forces Treaty
LNTS	League of Nations Treaty Series
NPR	Nuclear Posture Review (US)
NPT	Treaty on the Non-Proliferation of Nuclear Weapons
P5	Permanent five member states of the UN Security Council
PCIJ	Permanent Court of International Justice
SCR	Security Council resolution (UN)
TIAS	Treaties and Other International Acts Series (US)
TPNW	Treaty on the Prohibition of Nuclear Weapons
UK	United Kingdom of Great Britain and Northern Ireland
UN	United Nations
UNIDIR	UN Institute for Disarmament Research
UNRIAA	UN Reports of International Arbitral Awards
UNTS	UN Treaty Series
US	United States of America
USC	United States Code
USSR	Union of Soviet Socialist Republics
UST	United States Treaty Series
VCLT	Vienna Convention on the Law of Treaties

Acknowledgments

In writing this book, I have benefited from many discussions with friends and colleagues since my time as Senior Advisor for Strategic Planning in the US Department of State's Bureau of International Security and Nonproliferation and in the Office of the Under Secretary for Arms Control and International Security. Highly constructive notes from three anonymous reviewers have further assisted me in refining the main points and, hopefully, mitigating what some readers might find questionable in others. Where an individual's input has contributed specifically to my observations or arguments, I note as much in the text. I confine myself here to mentioning institutions that have supplied a frame in which to test ideas that I hope this book will bring to a wider audience.

The National Institute for Public Policy (NIPP) and the Nonproliferation Policy Education Center (NPEC) both kindly invited me over the past several years to publish my initial thoughts on the future of the NPT and to participate in discussions with highly engaged and deeply informed interlocutors concerned with the fate of nuclear arms control. The George Mason University National Security Institute and the George Washington University School of Law both welcomed me as a visiting scholar during the time that I was writing this book, as the Lauterpacht Centre for International Law at the University of Cambridge continued to provide a congenial 'academic home.' Two law reviews—the *George Mason International Law Journal* and the *Journal of Transnational Law and Policy*—published a lawyer's account of the Nuclear Ban Treaty and of Article VI of the Nuclear Non-Proliferation Treaty, which stand available for readers of this book interested in my more granular examination of some of the chiefly legal points.

The arguments and recommendations of policy in this book are my own, and none of these necessarily reflect the view of any government, organization, or individual beside myself.

Finally, I thank Stephen Wenham and Zoe Forbes, editors at Bristol University Press, and their production team for helping bring this book to press.

Preface

Armed aggression and geopolitical rivalry are realities of today's international environment. Whether because of these realities or in spite of them, negotiation toward nuclear arms control has faded from the scene. My purpose in this short book is to suggest practical steps that the US and its allies might take to revive the negotiated approach to nuclear arms control that, in the past, allowed important gains in national security and international stability. I offer an argument for realists in particular to return to nuclear arms control.

As reasonable as it sounds to say that they need to negotiate if nuclear-armed countries are to achieve arms control, the negotiated approach to arms control has come under threat. A new disarmament treaty, the Treaty on the Prohibition of Nuclear Weapons (TPNW), attracts support in many quarters, and it calls for immediate elimination of nuclear weapons without negotiation. The US and like-minded countries need a good faith effort toward nuclear arms control, because, without it, unrealistic calls for unnegotiated disarmament will complicate the already-difficult search for sustainable defense and security.

An earlier generation of strategists had the insight that nuclear arms control, in the world as it exists, requires negotiation. In the NPT, the Treaty on the Non-Proliferation of Nuclear Weapons, realists fashioned a negotiated approach, supporting a search for verifiable limits on nuclear arms, and, by so doing, they ensured that nuclear-weapon states would preserve the opportunity to achieve progress toward the eventual disarmament that was, and remains, a shared goal. It is timely to consider how to renew the negotiated approach.

As much as it entails coming to the table to explore possible solutions, a pledge to negotiate also entails refraining from conduct that pre-judges the outcome of negotiations. China refuses to come to the table and, having embarked on a large-scale nuclear weapons buildup, pursues a strategic *fait accompli* before negotiations have even begun.

Meanwhile, engagement with the NPT in the US is at a low ebb. Important voices nevertheless continue to champion this cornerstone treaty. As the US and its nuclear-weapon allies enter a season of political transition, I think it timely to turn our minds to overcoming the impasse at which nuclear arms control now stands. I hope those who read this book at least gain an appreciation of the risks that we face if unrealistic promises of immediate disarmament gain purchase.

Thomas D. Grant
Saddell House, Argyll
July 17, 2024

Introduction: A Tale of Two Treaties

This book is not about the history of nuclear arms control. It is instead about its possible future. A page of history nevertheless helps us understand where nuclear arms control might go and to appreciate why, as the title of this book suggests, nuclear arms control at the moment is in peril.

On July 16, 1945 the US tested the first atomic bomb. On August 29, 1949 the USSR tested its first atomic bomb. Finding themselves in a world in which two antagonists possessed weapons of enormous destructive power, political leaders began to consider possibilities for negotiating arrangements that might place those weapons under agreed limitations and controls.[1] Progress in the first decade was unsteady.[2] Countries reached a modest milestone in 1957 when they created the International Atomic Energy Agency, the IAEA,[3] a multilateral organization intended to support the peaceful use of nuclear technology and, 'so far as it is able', to ensure that nuclear technology transferred to countries without nuclear weapons is used *only* for peaceful ends.[4] The US, the USSR, and the UK—the last having tested its first atomic bomb in 1952—at length agreed to stop nuclear tests in space, the atmosphere, and oceans.[5] However, the Partial Test Ban Treaty, which the US, the USSR, and the UK adopted in 1963, did not place any other limits on the nuclear arsenals of these three countries, and it placed no obligation whatsoever on other countries. As for the IAEA, its governing statute left it to individual states to decide how, and whether, they would engage with the new Agency. No legally binding instrument with widespread membership did anything in itself to limit countries' freedom to develop or acquire nuclear weapons. With France having tested its first atomic bomb in 1960, and China having received technical support from the USSR to expedite its own development efforts—China would test *its* first bomb in 1964—concern grew that the spread of nuclear weapons was only beginning. US President John F. Kennedy, in 1963, the year of the test ban treaty, apprehended that in another decade or so as many as 25 countries would have nuclear weapons.[6]

This was the setting in which the US and USSR embarked in earnest to negotiate a nuclear arms control treaty that might attract all countries to join it, that would stop further proliferation of nuclear weapons, and that would supply the framework for negotiating still further limits on nuclear weapons for those countries that already had them. The result, reached in 1968, was the Treaty on the Non-Proliferation of Nuclear Weapons—the NPT.[7] The NPT, now having 191 states parties, embraces all but a small handful of UN member states,[8] plus the Holy See. Its ban on the further spread of nuclear weapons is thus far-reaching. The NPT's obligatory negotiations provision, Article VI, requires that its parties pursue negotiations toward further limits on nuclear weapons and, in time, even conventional disarmament. The treaty's substantive scope is thus ambitious. Moreover, the NPT does not stop at generalities. It entwines the arms control regime it creates with the IAEA, ensuring that 'non-nuclear-weapon states,' if they pursue peaceful use of nuclear energy, do so under safeguards that the IAEA deploys its personnel and infrastructure to verify. The NPT, which remains in force to this day, rightly is known as the 'cornerstone' of nuclear arms control.[9]

With the conclusion of the tenth five-yearly NPT Review Conference in August 2022,[10] however, the NPT finds itself at risk. Exponents of nuclear deterrence view the NPT's obligation to pursue negotiations at best with misgivings. For them, NPT Article VI, which expresses that obligation, has always looked too much like the thin end of a wedge aimed to pry away nuclear deterrence from national security strategy. Meanwhile, disarmament activists have come to view the NPT's negotiations requirement as a dead end. As the activists now see it, the nuclear-weapon states[11] have too long neglected NPT Article VI and its more ambitious disarmament goal. In particular in the US, a country without the commitment of which the NPT has no future, political stakeholders who have little else in common converge in discontent over the NPT. The world over, diplomats and nuclear policy makers speak of 'the NPT in peril.'[12]

The evidence that all is not well includes not just what people are *saying* about the NPT. It includes what people are doing—or, more precisely, what they are not doing. Nuclear arms control negotiations have come to a standstill. The impasse places stress on the treaty in two distinct ways.

First, on a reading of the treaty that is legally flawed but to which many states subscribe, nuclear-weapon states' failure to negotiate arms control and disarmament frees non-nuclear-weapon states from the constraints to which they agreed when they joined the treaty. The NPT obliges the vast majority of its parties—the non-nuclear-weapon states—never to produce or acquire nuclear weapons; and the others—the five nuclear-weapon states—never to supply a nuclear weapon or the means to get one to a non-nuclear-weapon state. These provisions—under Articles II and I of the NPT, respectively—impart content to the central term of the treaty, non-proliferation. They are unconditional. Yet many non-nuclear-weapon states say that their commitment to non-proliferation applies only to the extent that nuclear-weapon states implement other provisions of the NPT, in particular Article IV, paragraph 2, under which all parties undertake to facilitate the sharing of benefits of peaceful nuclear energy, and Article VI, under which all parties are obliged to pursue negotiations toward ending the arms race and disarmament. Many non-nuclear-weapon states suggest that their obligation under NPT Article III to accept safeguards negotiated with the IAEA similarly relaxes or disappears if they judge the nuclear-weapon states to fall short of implementing Articles IV(2) and VI.

There is no doubt that, when states negotiated the text that eventually became the NPT, they bargained for preferred terms. The text that they adopted—the NPT—arguably reflects a *quid pro quo* between nuclear-weapon states and non-nuclear-weapon states, in particular a trade-off between non-proliferation on the one hand and a right to peaceful nuclear use and an expectation of ongoing efforts toward arms control and disarmament on the other.[13] However, to say that states engaged in a *quid pro quo* to arrive at an agreed text does not tell us what the agreed text says. The negotiation dynamics when parties drafted a treaty do not control the interpretation and implementation of the treaty. If they did, then every term of every treaty would be nothing more than a framework for continued contestation. A *quid pro quo* in the *history* of the NPT's drafting does not equate to conditionality of the terms expressing the bargain that the parties reached and that continue to bind them.[14] To interpret the terms, one must read the treaty.

Many non-nuclear-weapon states, however, understand the historical *quid pro quo* behind the NPT to entail a present-day give-and-take. On that understanding, the commitment to non-proliferation is conditional. In addition to the treaty's drafting history, states and commentators who insist that non-proliferation is conditional invoke the decision at the 1995 NPT Review and Extension Conference to extend the NPT on a permanent basis, and to the Final Documents of the 2000 and 2010 Review Conferences. But the review conferences were not treaty-amendment conferences. The most consequential instrument adopted in any of these NPT meetings was Decision 3 of the 1995 Conference, at which the parties agreed to extend the NPT indefinitely in accordance with Article X(2).[15] Decision 3 interpolates no conditional term into the NPT.[16] To read the Non-Proliferation Treaty as a *conditional* Non-Proliferation Treaty is untenable as treaty interpretation.

The terms of the NPT notwithstanding, the conditional reading has gained widespread purchase. Because many maintain that delay in the pursuit of disarmament justifies laxity toward non-proliferation and safeguards, the negotiations standstill at which the nuclear-weapon states have arrived threatens to unravel the NPT from within.

Second, the negotiations standstill further places stress on the NPT by driving support toward a very different treaty. The NPT, while certainly not the best of treaties, is far from the worst. A new treaty, the Treaty on the Prohibition of Nuclear Weapons—the TPNW—takes an idealist's approach toward disarmament—and one that should give security-minded policy thinkers pause before we rush to retire the NPT. The TPNW, concluded in July 2017 and entered into force in January 2021,[17] in contrast to the NPT requires every state party that had a nuclear weapon after the TPNW's date of conclusion to carry out the 'irreversible elimination' of nuclear weapons and nuclear-weapon programs.[18] No nuclear-weapon state participated in drafting the TPNW or has signed it, but the TPNW already attracts support in many countries that rely on the US nuclear deterrent—including in countries on which the *US* relies as trade partners and strategic allies.

The difficulty with the TPNW is not that it reflects an aspiration to achieve a nuclear-weapon-free world. The difficulty with the TPNW is that it confuses the aspiration with reality. Moreover, its premise that the world is ready to disarm finds a receptive audience at a time when the waning engagement of policy makers with NPT Article VI fosters the impression that the incrementalism of good faith negotiations has failed. The story of the arms race and disarmament, once told in terms of the NPT, is now a tale of two treaties[19]—the NPT and the TPNW. Even if we dispense with the superlative degree of comparison (the labeling of 'bests' and 'worsts'), the virtues and the faults of the former are to be appreciated in the sobering light which the coming to life of the latter now affords. In short, the less promising the negotiations provision of the NPT, the more attractive, at least to key constituencies in many countries, the fast-track disarmament premise of the TPNW. Policy makers who care about strategic stability and security need to consider how the new treaty affects the old.

Policy makers also need to consider China's buildup of nuclear arms, for this affects both treaties and imparts new urgency to reviving nuclear arms control. China has embarked on the largest buildup of nuclear arms—and of the material and infrastructure to make nuclear arms—since the end of the Cold War. Arms racing remains with us even if only one country is racing, and, thus, the obligation under NPT Article VI to negotiate toward cessation of the arms race should not disappear from policy makers' agenda.[20] In view of the threat to stability and security that China's buildup presents, the negotiated approach to arms control is the realistic approach. The difficulty is that the TPNW, with its non-negotiable directive to disarm, continues to gain support. The TPNW will gain still more support, if the US and its allies answer China's buildup *only* with a countervailing buildup of our own. And in a disarmament discourse in which the TPNW prevails, we will face diplomatic, rhetorical, and political challenges as vexing as any of the Cold War.

The TPNW seems likely to remain a feature of the landscape of arms control and disarmament. Whether it comes to take over the landscape depends on the fate of the NPT.

The time has come to consider how we might reinvigorate the NPT, in particular by engaging afresh with the obligation under its Article VI to pursue and negotiate nuclear arms control. The stakes—in a world of burgeoning support for a prohibitionist alternative—are too high for us to ignore.

★ ★ ★

This book proceeds in five chapters.

Chapter 1 addresses the internal structure of the NPT, in particular the problematic—but widely accepted—reading that non-proliferation is a *conditional* obligation subject to present-day bargaining. On the conditional reading of non-proliferation, certain non-nuclear-weapon states argue that their Article IV 'inalienable' right to peaceful use allows them to hold materials and capabilities up to the very threshold of nuclear weapons. The conditional reading ignores the non-nuclear-weapon states' commitment to non-proliferation and safeguards. It threatens to defeat the purpose of the NPT. Its prevalence in international politics, however, has deprived the nuclear-weapon states of leverage in an era in which arms control negotiations are at an impasse and no new arms limitations are in prospect. One risk to the NPT thus comes from within: failing a revival of arms control negotiations in line with NPT Article VI, non-nuclear-weapon states, invoking Article IV, will be emboldened to pursue *any* nuclear capability short of an actually functioning weapon.

The other risk that I address in this book comes from without; namely, from a new treaty addressing nuclear arms. The TPNW obliges its parties to relinquish all nuclear weapons, if they have them, but it takes no account of the geopolitical realities that stand in the way of nuclear-armed competitors disarming. Failing to acknowledge that nuclear-armed competitors will not achieve disarmament unless they *negotiate* disarmament, the TPNW fails to supply a realistic path to arms control. Unsurprisingly, no state that has nuclear weapons yet has signed the TPNW. Chapter 2 argues that the TPNW is on a trajectory, however, to supplant the NPT as disarmament activists' main point of reference. If the TPNW were to fill a vacuum created by the fading of the NPT, then security diplomacy for the US and its allies and like-minded countries would face an uphill battle.

The figurative 'orphaning' of Article VI, the NPT's negotiations article, entails a risk that defense-minded policy makers should take seriously. It aggravates the confusion between historical *quid pro quo* and present-day interpretation and thus obscures the defect in non-nuclear-weapon states' claim that their commitment to non-proliferation is conditional; and it energizes support for an unrealistic alternative treaty, the TPNW.

Chapter 1 having considered how neglect of the obligation to pursue negotiations threatens both to erode the non-proliferation and safeguards provisions of the NPT and to sideline the NPT as a whole, and Chapter 2 having considered the TPNW and its defects, Chapter 3 takes a closer look at Article VI of the NPT. Diplomatic practice and international law are familiar with obligations to negotiate. The 'pursue negotiations' provision of Article VI expresses an obligation to negotiate. The obligation, as expressed, applies to all states parties to the NPT. As a practical matter, the nuclear-weapon states are the state parties that Article VI especially concerns. Though the US and USSR were the states parties holding the largest nuclear arsenals when the NPT was drafted and when it entered into force, the subsiding of their rivalry and the massive reductions in nuclear arms that they carried out did not change the meaning of Article VI. The 'arms race' limb of Article VI continued to apply.

The 'arms race' limb has special salience today. Chapter 4 turns to China's nuclear weapons buildup. Chapter 4 argues that China's buildup is incompatible with NPT Article VI, because a good faith obligation to pursue negotiations entails both that a party come to the negotiating table, which China has refused to do, and that a party refrain from conduct that either worsens the situation that it has committed to negotiate to resolve or imposes a *fait accompli* on other negotiating parties: China's buildup both worsens the situation and seems aimed at imposing a *fait accompli*. To respond adequately to China's nuclear weapons buildup, the US unfortunately will have to take material steps to enhance its deterrent capability. However, a well-considered use of diplomatic resources also should form part of a US response. In particular, the US should re-visit NPT Article VI and raise questions in appropriate settings about China's failure to negotiate.

Some countries have begun to take steps to hold China to account for its nuclear weapons buildup, but the US and its allies should do more, in particular with reference to Article VI. A quantitative nuclear arms race is undesirable, because it will introduce new risks and uncertainties to geopolitics, as well as impose significant financial costs. Advancing sound arguments and putting forward new proposals for negotiation in nuclear diplomacy are steps that the US and its allies should entertain.

Chapter 5 considers 12 specific topics that the US might seek to place on the negotiating agenda with a view to revitalizing Article VI of the NPT.

★ ★ ★

My argument in this book is that the fading of the NPT, if we do not arrest it, will pose risks that are of such magnitude that we should consider bold steps to breathe new life into that treaty. Disarmament activism, which found expression in treaty form with the adoption in 2017 of the TPNW, affects open societies much more than it affects our geopolitical challengers. The TPNW now attracts support from civil society and even governments in many countries. Support for the TPNW will grow, if civil society and key decision-makers perceive the NPT to have ceased to bring nuclear-weapon states to the negotiating table. A revitalized agenda for negotiating nuclear arms control thus would serve not merely to satisfy the formal requirements of a treaty clause. It would move the center of gravity in security diplomacy back toward the NPT and thus serve a strategic purpose, reassuring societies and governments that the cornerstone arms control treaty continues to supply the realistic framework for progress toward a safer world.

ONE

Three Pillars or One Foundation?

Before negotiations began on a nuclear arms control treaty, it appeared that a score or more countries soon would have nuclear weapons.[1] This was the nuclear future that the US Government National Intelligence Estimates forecast in the 1950s and to which President Kennedy later alluded.[2] The US and USSR, though for different reasons, agreed that far-reaching proliferation would undermine security, and, so, in the 1960s the superpowers started negotiations on nuclear arms control. The Treaty on the Non-Proliferation of Nuclear Weapons—the NPT—was the result. Historians are unlikely ever to settle the question whether the NPT is the reason many states that might have pursued nuclear weapons did not, but the fact remains: far fewer states came to hold nuclear weapons after conclusion of the NPT than experts had warned would do in the estimates they had ventured *before* conclusion of the NPT. Proliferation did not stop after 1968, but a world of 20 or more nuclear-armed rivals did not emerge.

Whatever its precise role in reducing the spread of nuclear weapons, the NPT undoubtedly was animated by a desire to achieve that end. Crucial was that practically all states join the NPT.[3] A treaty that only the states holding nuclear weapons joined would have left the door open to others lawfully to pursue nuclear weapons. But how to convince others to accept a new limit on their sovereign right to fashion a national defense?[4] One argument was that the envisaged limit—an obligation not to acquire nuclear weapons—would enhance international security and thus supply a self-sufficient reason for non-nuclear-weapon states to join the treaty.[5] At the same time, the drafters reckoned that an eventual non-proliferation treaty, if it were to attract the many states whom it aimed to prevent from having nuclear weapons, must accommodate certain other interests. The interest of non-nuclear-weapon states in the peaceful use of nuclear energy was foremost among these. In addition, it was a professed goal to achieve a more thorough-going arms control. Thus, as eventually drafted, the NPT addressed not only non-proliferation, but two further matters as well: peaceful use of nuclear energy by non-nuclear-weapon states and eventual nuclear and general disarmament.

Describing the NPT as a treaty of 'three pillars,' rather than one founded on the premise that all are served by the non-proliferation of nuclear weapons, has given rise to difficulty. On the 'three pillars' view, non-proliferation has no priority in the treaty, but, instead, vies for position through an ongoing process of bargain-seeking.[6] In particular, non-nuclear-weapon states seek flexibility in the definition of 'peaceful use' and leniency in the content and implementation of the safeguards that the NPT requires non-nuclear-weapon states to observe when they engage in peaceful use. Simultaneously, they insist that the nuclear-weapon states vigorously pursue disarmament. Let us start by considering peaceful use and safeguards, and then turn to the problem that arises, if the foundation of the NPT as a non-proliferation instrument is replaced by a tripartite give-and-take.

Peaceful use and safeguards

To recall, the animating purpose of the NPT is embodied in its first two articles. Article I obliges every nuclear-weapon state not to supply a nuclear weapon or the means to get one to any non-nuclear-weapon state; and Article II obliges every non-nuclear-weapon state not to possess or pursue a nuclear weapon. The further matters that the NPT addresses—peaceful use and disarmament—are addressed in Articles III and IV, and in Article

VI, respectively. Article IV ensures the non-nuclear-weapon states' right to peaceful use, and Article III requires any state engaging in peaceful use to observe IAEA safeguards. It is convenient to start by considering those two provisions in that order.

Article IV: peaceful use in the non-proliferation frame

NPT Article IV recognizes as 'inalienable' a right 'of all the Parties to the Treaty to develop research, production and use of nuclear energy for *peaceful purposes* without discrimination' (emphasis added). Article IV assures the non-nuclear-weapon states that the nuclear-weapon states will not retain the peaceful use of nuclear energy exclusively for themselves. Article IV is clear that this assurance is not absolute, for it expressly states that the 'inalienable' right that it recognizes may be exercised only 'in conformity with Articles I and II,' the non-proliferation provisions. Even if Article IV had not expressly stated the limits, these still would have followed on basic principles of treaty interpretation. Articles I and II, read together, bind all states parties, and without qualification.

For non-nuclear-weapon parties, Article IV is also, by necessary implication, subject to Article III.[7] Under Article III, each non-nuclear-weapon party is to negotiate and conclude a safeguards agreement with the IAEA. The right in Article IV to peaceful use of nuclear energy, while 'inalienable,' meaning that it cannot be taken away, is thus limited. The right operates subject to non-proliferation, which is a legal duty on all states, and for the non-nuclear-weapon states within further institutional-legal constraints laid down in Article III.

Article III: safeguards against proliferation

NPT Article III establishes an obligation on each non-nuclear-weapon state party to negotiate and conclude a safeguards agreement with the International Atomic Energy Agency, the IAEA.[8] The IAEA, an intergovernmental organization established in 1957, which is to say over a decade before the conclusion of the NPT, inspects facilities in non-nuclear-weapon states, including nuclear power plants, with a view to verifying that the states comply with the safeguards agreements they have concluded with the IAEA.[9] The IAEA performs verification functions, not enforcement, and its immediate concern is with verification that a state is observing its safeguards agreement with the IAEA.

The IAEA is not an NPT organ as such, much less an NPT compliance organ.[10] However, ever since the NPT Depository Governments notified the IAEA in March 1970 of the NPT's entry into force,[11] the Agency and Treaty have existed in close connection. On June 1, 1972 the IAEA Board of Governors approved INFCIRC/153—titled *The Structure and Content of Agreements Between the Agency and States Required in Connection with the Treaty on the Non-Proliferation of Nuclear Weapons*.[12] INFCIRC/153 remains the frame for comprehensive safeguards agreements.[13] In May 1997, spurred by the crisis over Iraq's safeguards evasion,[14] the IAEA Board introduced a Model Additional Protocol providing for further verification procedures.[15] At recent count, comprehensive safeguards agreements are in force for 182 states;[16] additional protocols, for 142 states.[17] Discovery by the IAEA of non-compliance with a safeguards agreement, in particular where the Agency has detected diversion of fissile material, such as plutonium or enriched uranium, raises proliferation concerns and, thus, can serve as the starting point for states to respond with measures intended to protect the shared interest in non-proliferation that the NPT enshrines.[18] While separate from the NPT, the IAEA and its safeguards system play an important role under NPT Article III.

NPT Article III is detailed and lengthy, running to 404 words in its English-language version. Paragraph 1 of Article III obliges as follows:

> [e]ach non-nuclear-weapon State Party ... to accept safeguards, as set forth in an agreement to be negotiated

> and concluded [with the IAEA in accordance with the IAEA Statute] and the Agency's safeguards system, for the exclusive purpose of verification of the fulfilment of its obligations assumed under [the NPT] with a view to preventing diversion of nuclear energy from peaceful uses to nuclear weapons or other nuclear explosive devices.[19]

As the text makes clear, a non-nuclear-weapon state party's legal obligations include both the specific terms of the agreement concluded with the IAEA—a safeguards agreement[20]—and the 'safeguards system' overall.

The UN Security Council, by resolution 1887 (2009) addressing the NPT and nuclear non-proliferation, '*[a]ffirmed* that effective IAEA safeguards are essential to prevent nuclear proliferation and to facilitate cooperation in the field of peaceful uses of nuclear energy.'[21] The first clause in this paragraph of Security Council resolution 1887 affirms the limitation that non-nuclear-weapon states accepted when they became parties to the NPT; the second clause recalls that they also retained the liberty to pursue peaceful use of the atom—in particular, for purposes of generating energy. The form of words expressly connects safeguards to the foundational goal of the NPT—non-proliferation—while recalling that non-nuclear-weapon states obtained an assurance of access to peaceful use of nuclear energy when they agreed to participate in this organized legal regime for non-proliferation. The IAEA safeguards system, in this double-valent way, has been central to the NPT since the NPT entered into force.[22]

Shaky pillars and the peaceful use conceit

On the view that the NPT is a treaty of 'three pillars,' non-proliferation, peaceful use of nuclear energy, and disarmament are said to stand in mutual symmetry, no one bearing a greater weight than another when states come to interpret and apply the treaty. Not found in the NPT itself, the expression 'three pillars' is one that states were using by the time of the 1985 NPT Review Conference.[23] The expression is now ubiquitous in policy statements and academic literature.[24] The US State Department currently uses the expression,[25] and the UK Foreign, Commonwealth and Development Office uses one much like it.[26] Daniel Joyner, pre-eminent exponent in academia of the 'three pillars' interpretation, says that the 'pillars' are of such 'principled' importance to the NPT, that they '*together* comprise the object and purpose of the NPT.'[27] As I have suggested, however, there are reasons to be cautious about describing the NPT this way.

For some time, under the 'three pillars' view, non-nuclear-weapon states have interpreted the recognition of their right to peaceful use of nuclear energy as a bargaining chip. Instead of a fixed and circumscribed right, peaceful use, under the 'three pillars' view,' has been invoked to extract concessions that the treaty does not provide. Of most serious concern in this regard is a supposed right to acquire materials and technologies that bring a state to the cusp of nuclear weapons. Speaking as Special Representative of the President of the United States for Non-Proliferation of Nuclear Weapons, in 2004 Ambassador Jackie W. Sanders said:

> The central bargain of the NPT is that if non-nuclear-weapon states renounce the pursuit of nuclear weapons, and comply fully with this commitment, they may gain assistance under Article IV of the Treaty to develop peaceful nuclear programs ... However, Parties cannot afford to ignore the fact that several countries have exploited Article IV to advance their illicit nuclear weapons programs and threaten international security. These countries have not lived up to their end of the bargain, and if we allow this abuse to continue, the net-value of peaceful nuclear cooperation will diminish, and the security benefits derived from the NPT will erode.[28]

Ambassador Sanders here referred to the bargain that the NPT parties reached when entering the treaty; that is to say, the fixed terms expressed in the treaty and binding them all. As Ambassador Sanders observed, the NPT is a non-proliferation treaty. Article IV, with its promise of assistance for peaceful nuclear use, was a further inducement to non-nuclear-weapon states to participate in the NPT—and thus to accept the obligation to forego nuclear weapons. To say, instead, that Article IV permits a non-nuclear-weapon state to advance a weapons program, limiting the state only from taking the final steps to a functioning weapon, is an interpretive conceit. Peaceful use overruns Articles I and II, and loses its effective meaning, if it allows all but proliferation.

The Under Secretary of State for Arms Control and International Security, John Bolton at the time, agreed with Ambassador Sanders. According to Bolton:

> The central bargain of the NPT is that if non-nuclear weapons states renounce the pursuit of nuclear weapons, they may gain assistance in developing civilian nuclear power. This bargain is clearly set forth in Article IV of the Treaty, which states that the Treaty's 'right' to develop peaceful nuclear energy is clearly conditioned upon parties complying with Treaty Articles I & II. If a non-nuclear weapon state party seeks to acquire nuclear weapons and thus fails to conform with Article II, then under the Treaty that party forfeits its right to assistance in the development of peaceful nuclear energy.[29]

The text of the NPT supports the interpretation that these officials expressed. The peaceful use provision—Article IV—expressly incorporates the non-proliferation provisions, Articles I and II. Together, Articles IV and II define the bargain that the non-nuclear-weapon states accepted: Article IV is conditional upon their compliance with Article II.

Considering the political economy when the NPT was drafted, the interpretation is all the more convincing that the 'central bargain' was the acceptance by the non-nuclear-weapon states of an unconditional legal obligation to refrain from future weapons development, exchanged for a *conditional* recognition from all states of a right to future peaceful development. A small handful of states were uniquely positioned either to supply, or to deny, access to the technology and material needed for peaceful nuclear use. Considering that these states had the requisite solidarity to draft and adopt the NPT, they might have pursued a policy of nuclear exclusivity—not only in the realm of weapons but also of peaceful use. The non-nuclear-weapon states, by accepting the NPT's non-proliferation obligations, gained an assurance—conditioned on the faithful observance of non-proliferation—that the nuclear-weapon states would not stifle their access to the benefits of peaceful use of nuclear energy.

An idealist might say that the nuclear-weapon states conceded nothing to the non-nuclear-weapon states when adopting the NPT, because (the idealist would say) a general right to peaceful development exists whether or not states acknowledge it.[30] But, whether or not such a general right extends to nuclear energy, there is no general *obligation* to aid a state. Article IV(2), obliging all parties to 'facilitate' peaceful use of nuclear energy, does not codify an existing obligation. It stipulates a new one. States at the time of the NPT's drafting recognized that Article IV was an important concession by the nuclear-weapon states.[31]

And, even if an obligation to facilitate peaceful use already had existed, there is still the reality of self-help in international relations. In France, one of the nuclear-weapon states, foreign ministry legal advisers a few months before conclusion of the NPT expressed concern that, misconstrued, guarantees of non-proliferation even would 'run a risk of justifying the launch of a preventive war.'[32] In a matter as fraught with strategic consequences as nuclear proliferation, non-nuclear-weapon states knew full well that nuclear-weapon states could not be counted upon to tolerate the further spread of nuclear weapons or of the technology necessary to their manufacture. In Article IV itself, a bargain invaluable to non-nuclear-weapon states inheres. Under the agreed terms that bargain is not open-ended.

One former nuclear non-proliferation planner of the US Joint Chiefs of Staff refers to Article IV as '[p]erhaps the NPT's most controversial provision.'[33] The keenest controversy arises from the contention that the assurance of peaceful use equates to a right to possess any material or technology short of an actually functioning nuclear weapon. In October 2006, Mohamed ElBaradei, as IAEA Director General, referred to 'virtual nuclear-weapon States'—countries that have nuclear fuel-making capability and for which 'conversion time [to nuclear weapons-grade material] is very short.'[34] Negotiating does not change the material and technological requirements for building a nuclear weapon. Past a point, possessing certain material and technology is tantamount to possessing a nuclear weapon.

Albert Wohlstetter observed nearly a half century ago that the NPT is a '*nonproliferation* treaty, not a nuclear development treaty.'[35] But one need not resort to a teleological approach to treaty interpretation to recognize that Article IV does not allow a non-nuclear-weapon state to make itself a 'virtual nuclear-weapon State'. The overall nature of the NPT is well reflected in the specific text of the NPT. The Article IV guarantee of peaceful use is expressly limited. The guarantee is limited by the unconditional commitment to non-proliferation contained in NPT Articles I and II; and it is limited by the safeguards that all non-nuclear-weapon states parties must negotiate and observe in accordance with NPT Article III. As one former US nuclear arms control official has said, the 'safeguardability approach reconciles the text of Article IV with the rest of the Treaty, with its negotiating record, and with long-standing international approaches to nuclear technology.'[36]

In the NPT as drafted, non-proliferation is absolute and peaceful use is qualified. Subsequent practice demonstrates the wisdom of this arrangement. The Democratic Republic of Korea (North Korea) tested its first nuclear weapon in 2006, which was four years after it had discontinued cooperation with IAEA inspectors.[37] Where there is breakdown of Article III and of the safeguards system, abuse of the Article IV right to peaceful nuclear development naturally follows. To read Article IV as a guarantee of access to any material or technology short of a functioning weapon is incompatible with non-proliferation.

Again, the states with nuclear weapons in the 1960s had choices. Whether they regularized non-proliferation under a multilateral treaty, or left the matter in the less predictable domain of *ad hoc* policy making and self-help, was a question of considerable moment for states that did not have nuclear weapons. The prospect of crushingly expensive nuclear arms races with neighbors and regional competitors, and the security risks of multiple nuclear dyads among heretofore non-nuclear-weapon states, also gave those states reason to join a non-proliferation treaty.[38] The choice of a legalized approach, placing non-proliferation in the frame of a treaty, in itself entailed considerable benefits for the non-nuclear-weapon states. It is a conceit to read Article IV peaceful use as a 'pillar' equal to Article II non-proliferation, all the worse to read Article IV as a sliding scale along which peaceful use includes everything up to the threshold of nuclear weapons. The better reading is that which places peaceful use in its proper perspective—a qualified assurance offered in order to reinforce non-proliferation, the one foundation on which the NPT stands.

★ ★ ★

The persuasiveness of the better reading, however, does not in itself change the politics in which the NPT operates. Because so many commentators and states accept the 'three pillars' view, an understanding is prevalent that the liberty recognized in Article IV and the safeguards required under Article III are elastic. On that understanding, the former enlarges and the latter dwindle, depending on the conduct of the nuclear-weapon states.[39]

One way to attempt to rescue the NPT from the 'all but' weapons conceit is to continue to articulate the better argument: the NPT is a *non-proliferation* treaty, not a peaceful-use-facilitating treaty. The difficulty with leaving it at that—counting on the better argument—is that the better argument, in many quarters, already has lost. This brings us to a different approach—reviving negotiations for nuclear arms control.

Article VI: pursuit of negotiations

Under the prevalent, though flawed, understanding of the NPT as a treaty of 'three pillars'—non-proliferation, peaceful use, and disarmament—the commitment of the non-nuclear-weapon states to remain so is conditional not only upon the availability to them of peaceful nuclear energy; it is conditional as well upon the nuclear-weapon states pursuing negotiations toward effective measures on arms control and disarmament. From a jurist's standpoint, there are difficulties with characterizing the NPT as a 'synallagmatic' agreement—one containing terms that a party is obliged to implement only when another party implements other terms in the treaty.[40] The stability of treaty relations is one of the core principles of international law,[41] and the grounds for ceasing to observe a treaty are narrow.[42]

Moreover, as I recalled earlier in this chapter, the text of the NPT is clear where it identifies the Article IV assurance regarding peaceful use as conditional upon the non-nuclear-weapon states' performance of their Article II obligation of non-proliferation. The conditioning of Article IV is explicit, and Article III safeguards strengthen it. No such conditioning qualifies Article II. The drafting history records that some states proposed conditioning Article II on the fulfillment of Article VI.[43]

Conditionality between *those* provisions was rejected. Thus, the adoption of Article II in unconditional form was deliberate and considered.

However, as a matter of the *politics* of the NPT, conditionality remains in the discourse.[44] The text and drafting history notwithstanding, many non-nuclear-weapon states say that non-proliferation depends on the pursuit of negotiations toward arms control.[45] Their position is a political reality that policy makers must take into account.

Because many parties declare Article VI—the negotiations provision—a coequal 'pillar' of the NPT, pursuing negotiations would give nuclear-weapon states renewed diplomatic leverage to call for faithful observance of the NPT's foundational rule, non-proliferation. This is not the only reason of NPT politics to consider resuscitating nuclear arms control negotiations, but it is one. Before turning to the other essentially *political* reason for a return to negotiations—the rise of the nuclear ban treaty, the TPNW—let us consider the logic of the NPT negotiations provision as the self-contained *legal* text that it is.

Reasons to negotiate, negative and affirmative

A succinct text, but later freighted with difficulties, NPT Article VI provides as follows: 'Each of the Parties to the Treaty undertakes to pursue negotiations in good faith on effective measures relating to cessation of the nuclear arms race at an early date and to nuclear disarmament, and on a treaty on general and complete disarmament under strict and effective international control.'[46]

Self-evidently, Article VI obliges the parties to pursue negotiations. If the parties do as Article VI obliges, then they achieve one essentially negative, yet vital, object: they avert a future in which there is no possible end to arms racing and no possibility of disarmament.[47] Ensuring that the door is never closed to a world without nuclear weapons is a seemingly unambitious object, yet the realities of strategic competition make it no foregone conclusion that nuclear-weapon states will leave that door open. Article VI performs this conservatory function: it reduces the risk that arms control and disarmament will disappear from the nuclear-weapon states' agenda.

Article VI performs an *affirmative* function, too. By obliging them to pursue negotiations on effective measures relating to cessation of the arms race and to nuclear disarmament, Article VI leads the parties to develop higher levels of trust and to explore practical mechanisms under which disarmament becomes a stronger possibility. The affirmative function of Article VI finds reflection in the NPT preamble, which expresses the states parties' desire 'to further the easing of international tension and the strengthening of trust between States in order to facilitate the cessation of the manufacture of nuclear weapons.' If one were to identify a single most valuable insight in the NPT, then that would be that the challenge of the nuclear-weapons age is *not* to reach agreement that the world would be better without nuclear weapons; but, instead, to achieve the trust and practical mechanisms needed to reduce weapons holdings until none remain. The desire for a nuclear-weapon-free world is widely shared and was from the start of the NPT. Agreed pathways to that goal are what elude us—and are what Article VI obliges countries to seek.

One measure of the NPT's current purchase on US policy thinking is the periodic Nuclear Posture Review—the NPR. The NPR, in accordance with legislative mandate,[48] articulates US nuclear policy, strategy, capabilities, and force posture on a five- to ten-year timeline. The most recent NPR was transmitted to Congress in classified form on March 28, 2022;[49] the 2022 NPR was released in unclassified form on October 27, 2022.[50]

The 2022 NPR states the affirmative case for the NPT. It posits that the NPT 'has made the world safer and more prosperous' and that *all* NPT parties 'continue to benefit from the Treaty.'[51] According to the 2022 NPR, the NPT advances peaceful uses of nuclear energy.[52] Leaving it at that, however, says both too much and too little. It says too much, in its potential implication that peaceful use constitutes a coequal 'pillar' of the NPT. It says too little, in its failure to make a complete case for the NPT. The NPT is not only about achieving good outcomes. It is also—originally, *mainly*—about avoiding worst outcomes.

This aspect of the NPT—avoiding the worst—is reflected in one of the treaty's central features: the NPT reminds its parties that nuclear weapons remain a geopolitical reality. There would have been no use in the NPT recognizing the existence of the five nuclear-weapon states, or in obliging them to pursue negotiations on arms control and disarmament, if it were otherwise. The NPT is not, as some non-nuclear-weapon states would have it, a peaceful use treaty; nor is it, as the disarmament community would have it, a rapid disarmament treaty. It is, instead, a treaty to prevent worst-case outcomes, in which destabilizing arms buildups upset the calculus of deterrence and introduce hard-to-manage risks—and in which recalcitrant nuclear-weapon states eschew even the endeavor toward a safer geopolitical settlement. To understand the NPT in the round, one needs to recognize this negative dimension—an assurance that arms control will not veer toward self-defeating absolutism and unrealistic timelines—as much as the affirmative dimension—an assurance that the parties will pursue negotiations toward meaningful, practical nuclear controls and restraints and eventual disarmament.

Overreading Article VI

It is important that we not overread Article VI. Article VI does not introduce specific obligations to reduce nuclear arms. And it does not create a legal obligation to bargain about the aid states share to facilitate peaceful use. Article VI is a self-contained provision on its terms and by the logic that animated its adoption.

Ambitious but flawed readings of Article VI have vexed the NPT practically from the start. Even as cerebral a reader as Joseph Nye, Jr., writing in 1985, referred, erroneously, to 'arms reductions called for [*sic*] in Article VI.'[53] As for states parties, many among them long have read Article VI to support their attempts to enlarge the scope of Article IV peaceful use and diminish the scope of Article III safeguards.

A wellspring of particular confusion over Article VI is the 1996 Advisory Opinion of the International Court of Justice (ICJ) on *Threat or Use of Nuclear Weapons*. According to the ICJ: 'The legal import of that obligation [to pursue negotiations] goes beyond that of a mere obligation of conduct; the obligation involved here [under Article VI] is an obligation to achieve a precise result—nuclear disarmament in all its aspects.'[54]

This is a strange way to describe an obligation to negotiate. In diverse circumstances, negotiation clauses in treaties have *not* meant an obligation to reach a precise result.[55] They have meant what Article VI self-evidently means: an obligation to *endeavor* toward a result. With little explanation for its departure from a long-standing and consistent jurisprudence about negotiation (which I will consider further in Chapter 3), the ICJ espoused the idea, however, that NPT Article VI lays down an obligation on all states to disarm. This act of judicial transformation is all the more perplexing, when we recall that the UN General Assembly had not asked the ICJ for an interpretive opinion in regard to Article VI.[56]

Skepticism about the ICJ's pronouncement that Article VI obliges disarmament as a *result* (and, thus, is unlike other negotiations clauses, which oblige the negotiated endeavor *toward* a result) is hardly confined to hawkish corners of the nuclear strategy community. The skepticism is widespread. Marco Roscini, who now holds the Swiss Chair of International Humanitarian Law at the Geneva Academy of International Humanitarian Law and Human Rights,[57] wrote in 2015 about the interpretation of Article VI and the puzzlement that the

1996 Advisory Opinion continues to raise. Professor Roscini's observations merit reciting:

> The ordinary meaning of Article VI of the NPT does not suggest an obligation to bring the negotiations to a successful conclusion, for instance by adopting a treaty on nuclear disarmament, but only 'to pursue negotiations' in good faith. This is confirmed by an interpretation that takes account of the context, in particular of the aspirational language of the preambular paragraph [of the NPT] declaring the intention of the parties 'to achieve at the earliest possible date the cessation of the nuclear arms race and to undertake effective measures *in the direction of* nuclear disarmament.' [emphasis Roscini's] Compare the vague language of Article VI to the far more specific text of Article III(4) ... which provides, *inter alia*, for the obligation of the non-nuclear weapon states to negotiate *and* conclude safeguards agreements, with specified characteristics, with the ... [IAEA]. Finally, the *travaux préparatoires* ... confirm that it was possible to include Article VI in the final text ... exactly because it did not entail a commitment to conclude a treaty on nuclear disarmament.[58]

The 'aspirational language' in the NPT's preamble that Roscini notes indeed is precisely that—a statement of a directional goal—and it confirms the meaning that follows naturally from Article VI itself. The directional interpretation also accords with the superpowers' preferences at the time of drafting.

Sometimes said to have transformed Article VI into an obligation of disarmament result, the 1995 NPT Review and Extension Conference in fact restated the existing terms of the treaty. It is unsurprising that the parties in that setting drew particular attention to the goal of disarmament. Tellingly, in the preamble to Decision 2—Principles and Objectives for Nuclear Non-Proliferation and Disarmament—where the Conference of Parties reiterated that disarmament is an 'ultimate goal [...]', the Conference also reiterated that the goal is to be pursued by 'continu[ing] to move with determination towards the full realization and effective implementation of the provisions of the Treaty.'[59] The directional aspect remained clear. Moreover, the specific arms control steps that Decision 2 lists (in paragraph 4(a), (b), and (c))[60] are a 'programme of action,' not a modification of the NPT to embrace new obligatory terms. Article VI remains an obligation to pursue the goal of disarmament through negotiation, not an obligation to disarm.

And, yet, the disarmament community, overreading Article VI, has portrayed it for decades as precisely that—an obligation to disarm. Observing the nuclear-weapon states not to have disarmed, the disarmament community judges the NPT inadequate or an outright failure. Realists overread Article VI as well and too readily accept the description of the NPT as a treaty of 'three pillars.' The realists judge the NPT an idealist's distraction or a threat to sound strategy. The reading of Article VI as an obligation to disarm and the practice of calling it a 'pillar' coequal to non-proliferation have led disarmament activists and realists alike, each for their own reasons, to resile from the NPT.

★★★

Matthew Costlow, who served as a Special Assistant in the Office of Nuclear and Missile Defense Policy at the US Department of Defense from 2019 to 2021, observes that '[s]elf-styled "realists" have largely abandoned [study of] ... the post-Cold War arms control environment, with the regrettable side effect of allowing an echo chamber to form among those who already favor arms control—only debating amongst themselves the scope and pace of disarmament, not its prudence.'[61] We should not only regret the disarmament 'echo chamber' among academic writers. We should be alarmed at the one-sided trend in diplomacy and statecraft around arms control as well. If NPT Article VI becomes a political orphan, then we will struggle to address the growing influence of a new and problematic alternative, the Treaty on the Prohibition of Nuclear Weapons, to which I now turn.

TWO

The TPNW Challenge

The NPT, long the 'cornerstone' of efforts by governments in their pursuit of nuclear arms control, was for nearly 50 years the only open multilateral treaty that embodied the aspiration to rid the world of nuclear weapons. The states that drafted the NPT recognized that the challenge is not to reach consensus about a goal; they already shared the goal of nuclear disarmament. The challenge was, and remains, instead, how to achieve the goal in a world where practical arms control steps are constrained by strategic realities. They drafted the NPT with the realities in mind.

Today, however, we live in a world of *two* treaties concerned with nuclear disarmament. The second treaty—the TPNW—now gathers adherents. More governments are signing the TPNW as time goes on. Of equal consequence or more, the disarmament community, which once placed its faith in the NPT, has shifted its focus to the TPNW. If, as Matthew Costlow suggested, we inhabit an 'echo chamber' in which realists, having little enthusiasm for the NPT, remain silent, and in which disarmament activists lead the NPT discourse, then now we are approaching the point where even the echo is starting to fade. The activists now promote the TPNW and have less and less to say about the 'cornerstone' that preceded it.

Let us consider the TPNW in further detail, including the circumstances that led parties to draft it and the provisions that distinguish it from the NPT.

Origins of the TPNW

The pre-history of the TPNW covers a long course of humanitarian treaty-making and disarmament politics.

Going back as far as the 1925 Geneva Protocol which prohibits 'asphyxiating, poisonous or other gasses' and 'bacteriological methods of warfare,'[1] states through the 20th century agreed prohibitions upon certain categories of weapon. Prohibitions of biological and chemical weapons, respectively under the 1972 Biological Weapons Convention (BWC)[2] and 1993 Chemical Weapons Convention (CWC),[3] are landmarks in this practice. More recent prohibitionary instruments include the 1997 Landmine Convention[4] and the 2008 Convention on Cluster Munitions.[5] These, and their forerunners, enliven the belief that far-reaching treaty measures might resolve the humanitarian challenges that warfare, particularly in its modern manifestation, entails. While the BWC and CWC enjoy wide subscription—185 and 193 states parties, respectively—the Landmine Convention and Cluster Munitions Convention have attracted less support and more opposition.[6] The US, Israel, Iran, India, Russia, and China are among the states that are parties neither to the Landmine Convention nor to the Cluster Munitions Convention.

The potential humanitarian impact of the use of nuclear weapons had motivated the requests for ICJ advisory opinions on nuclear weapons by the World Health Organization and UN General Assembly in the 1990s,[7] and disarmament activists had contemplated a general ban practically from the start of the nuclear age.[8] In the first decade of the 21st century, the activists directed their efforts toward a possible ban treaty. The UN Secretary-General in 2009, at the Test-Ban Treaty Conference, referred to 'new momentum for a world free of nuclear weapons.'[9] In the years after the 2010 Review Conference of the Parties to the NPT, the text that became the TPNW took shape.[10] Unlike the NPT, the TPNW declares an immediate ban on nuclear weapons.

Practically all states recognize that the use of a nuclear weapon in most circumstances would impose terrible human cost.[11] In 2010, at the NPT Review Conference, the states parties, in the Final Document of the Review Conference, recalled the risk that 'catastrophic humanitarian consequences . . . would result from the use of nuclear weapons'—a routine reminder.[12] It was after this that disarmament activists invoked the humanitarian concern justifiably highlighted by the NPT Review Conference and pursued a nuclear weapons ban.[13] Indeed, the conference that the UN General Assembly later convened outside the NPT framework to draft a nuclear weapons ban attributed the ban effort to 'recent discourse centred on promoting greater awareness and understanding of the humanitarian consequences' that the ban conference said 'would result from *any* use of nuclear weapons.'[14] According to the ban conference, 'renewed interest in the humanitarian impact of nuclear weapons was first manifested' in the 2010 Review Conference Final Document.[15]

The ban conference's description of 'renewed' interest in humanitarian impact is puzzling. The nuclear-weapon states that adhere to rule of law, and where democratic elections hold governments to account, have never ceased to be concerned about the humanitarian impact of nuclear weapons.[16] Recalling humanitarian concern at the 2010 NPT Review Conference was not to repair a lapse or break new ground, and, so, to say that it was in 2010 that humanitarian concern was 'first manifested' would be misleading.

It is a relative quibble to draw attention to the ban conference's doubtful account of disarmament history. A more serious fault was the silence at the ban conference in regard to strategic context. It was not in a vacuum that the NPT Review Conference in the Final Document recalled long-standing humanitarian concerns. The Review Conference also recalled that eventual nuclear disarmament must take place 'in a way that promotes international stability, peace and security, and based on the principle of undiminished and increased security for all.'[17] The NPT states parties thus recognized that steps toward nuclear disarmament will promote international stability, peace, and security only if states maintain due regard for the realities of the nuclear age. Most salient among those realities are the acknowledged possession of nuclear weapons by eight states, among them the five nuclear-weapon states of the NPT, and the pursuit of nuclear weapons by others.[18] Calling for disarmament, while ignoring the international environment in which states still threaten to use nuclear weapons and further states want to get them, is not an effective way to address the shared humanitarian concern over the consequences of the use of nuclear weapons.

The US, for its part, understands disarmament to be achievable only in the presence of 'strict and effective international control' over weapons in general, that being the express language of NPT Article VI.[19] The US understanding corresponds to the legal text and the geopolitical realities: the US, its nuclear allies, and the large parts of the world covered by alliance guarantees count on nuclear weapons to stabilize international relations.[20] Security against conventional armed aggression and other weapons of mass destruction also owe to the deterrent that the US and its nuclear allies maintain.

Notwithstanding the geopolitical realities—and the legal undertakings in the NPT that reflect them—the UN General Assembly in 2016 took its lead from the disarmament community.[21] The General Assembly convened the conference, the pronouncements of which about 'recent' humanitarian discourse I quoted, in order to produce a treaty that would ban nuclear weapons immediately, which is to say without the antecedent steps that states for half a century had recognized as essential.[22] On July 7, 2017, the ban conference adopted the TPNW.[23] The TPNW opened for signature on August 9, 2017 and entered into force on January 22, 2021, having obtained 50 ratifications, acceptances, approvals, or accessions.[24] Honduras was the 50th state to ratify.[25]

As of July 2024, the UN Office for Disarmament Affairs counted 93 states as TPNW signatories, of which the treaty was in force for 70, among them the Cook Islands and Niue, which are non-UN-member states in free association with New Zealand.[26]

The provisions of the TPNW

A short treaty, the TPNW contains a preamble and 20 articles, inclusive of concluding and formal provisions. Article 1 sets out the nuclear prohibitions, which are comprehensive. Article 1 bans, *inter alia*, the development, testing, production, manufacture, or any acquisition, possession, or stockpiling of nuclear weapons or other nuclear explosive devices.[27] Among other elements of Article 1 is a ban on any assistance, encouragement, or inducement to engage in any of the prohibited activity.[28] The prohibitions expressed in Article 1 of the TPNW call to mind NPT Articles I and II, which prohibit the development or acquisition of nuclear weapons and assistance for the same; except that in the TPNW the addressees are *all* the parties to the treaty.

The TPNW does acknowledge, particularly in its Article 4,[29] that states exist that hold nuclear weapons. No state that holds nuclear weapons participated in drafting the treaty, but the drafters envisage that states that hold nuclear weapons, in time, will become party. Article 4, paragraph 1, obliges '[e]ach State Party that after 7 July 2017 owned, possessed or controlled nuclear weapons or other nuclear explosive devices and eliminated its nuclear-weapon programme' to cooperate with 'the competent international authority' (to be designated by the states parties in accordance with Article 4, paragraph 6) for the purpose of verification.[30] No TPNW party or signatory to date belongs to the Article 4 category.[31]

Apparently acknowledging that the instantaneous disarmament of a nuclear-armed state is not practically feasible even if the state agrees to disarm, the TPNW's drafters included a 'notwithstanding' clause, Article 4, paragraph 2. Article 4, paragraph 2, merits setting out in full:

> Notwithstanding Article 1(a), each State Party that owns, possesses or controls nuclear weapons or other nuclear explosive devices shall immediately remove them from operational status, and destroy them as soon as possible but not later than a deadline to be determined by the first meeting of States Parties, in accordance with a legally binding, time-bound plan for the verified and irreversible elimination of that State Party's nuclear-weapon programme, including the elimination or irreversible conversion of all nuclear-weapons-related facilities. The State Party, no later than 60 days after the entry into force of this Treaty for that State Party, shall submit this plan to the States Parties or to a competent international authority designated by the States Parties. The plan shall then be negotiated with the competent international authority, which shall submit it to the subsequent meeting of States Parties or review conference, whichever comes first, for approval in accordance with its rules of procedure.

An initial observation about Article 4(2) is that it does nothing to change the legal position that results under Article 1. Any nuclear-weapon state that does, in future, decide to become a party to the TPNW by becoming party makes itself subject to the ban on nuclear weapons that Article 1 articulates. The Article 4(2) 'notwithstanding' clause does not qualify the ban. It implements it.

On the terms of the TPNW, implementing the ban would be a matter of technical modalities and procedure.

The first technical modality would be the removal of the state's nuclear weapons from 'operational status.' The intention behind the 'operational status' clause would appear to be to provide the nuclear-weapon state a grace period between accession to the TPNW and physical elimination of the state's nuclear weapons. Article 4(2) assumes that removal from operational status would be a unilateral step and that it would take no time at all ('immediately'). There are reasons to doubt whether the clause would work. As soon as the state has no nuclear weapons in 'operational status,' the state's strategic posture will be very different from what it was before. The TPNW requires that the nuclear-weapon state immediately place itself in a posture very much like that of a disarmed state. In present geopolitical reality, security planners will find little comfort (and no grace) in the distinction between 'irreversible elimination' (which the next clause of Article 4(2) requires) and immediate removal from operational status. The 'operational status' clause prescribes a technical modality to address a problem that is geopolitical and strategic, not technical.

After the immediate removal of the state's weapons from 'operational status,' the state is to 'destroy' its weapons 'as soon as possible but not later than a deadline to be determined.' The emphasis here, too, is on technical modality, and, again, it is misplaced. Destruction of a nuclear arsenal, or even part of one, is a daunting task, as the US and Russia can attest from the early post-Cold War experience. However, a great deal more is needed than the physical act of disarmament. An antecedent shift in international relations was necessary before the then-superpowers could pursue the physical act in earnest. The same would be needed today.

The concession in Article 4(2) to a later 'deadline,' read in context, serves one purpose: to afford the time necessary to implement the physical act of 'elimination of that State Party's nuclear-weapon programme, including the elimination or irreversible conversion of all nuclear-weapons-related facilities.' The deadline is 'to be determined' through a procedural mechanism, 'the first meeting of States Parties.' The deadline, on these terms, is a determination, not a negotiated outcome. It is hard to see a nuclear-weapon state, under current circumstances, accepting an intergovernmental body's determination about a matter as sensitive to the state's sovereign interests as the possession of nuclear weapons.

It is true that Article 4(2) refers to a 'negotiated' step. Article 4(2) envisages a negotiation in respect of a plan that the state shall implement for the elimination ('or irreversible conversion') of its nuclear weapons. So the 'negotiated' step concerns another technical modality—the plan by which the state shall eliminate its nuclear weapons. And the plan is negotiated only to a point. The negotiation is not with those states which likely will concern the disarming state the most—that is to say, the other nuclear-armed states.[32] Instead, it is to the other TPNW parties or 'to a competent international authority designated by the States Parties' that the nuclear-weapon state must submit the plan. The state must submit the plan 'no later than 60 days' after the TPNW enters into force for the state. Evidently, after it submits the plan, the state may negotiate about it with 'the competent international authority,' but such negotiating will not be for long either: the competent international authority 'shall submit [the plan] to the subsequent meeting of States Parties or review conference, whichever comes first, for approval in accordance with its rules of procedure.' It is not entirely clear from the text of Article 4(2) whether the states parties must give their 'approval' in consensus with the nuclear-weapon state that submitted the plan. A plausible interpretation of the text is that the states parties at this stage *prescribe* the 'legally binding, time-bound plan' for weapons elimination, even if a difference as to the plan's content or deadline subsists between the 'competent international authority' and the nuclear-weapon state. But, alternatively, it is plausible that, if the parties fail to agree to a plan, then the matter will fall into deadlock.[33] In any event, it does not appear that Article 4(2) contemplates a colloquy between the states parties as a whole and the nuclear-weapon state on any matter other than the technical modality of arms elimination. Nor does Article 4(2) provide for realistic timelines, the 60-day mark following entry into force being the mandatory time-limit for the nuclear-weapon state to submit its weapons-elimination plan, and either the next meeting of states parties or review conference—'whichever comes first'—being the time-limit for plan approval.

Even the reductions in nuclear armament by the US and Russia, which, though deep, were a far cry from elimination of either state's nuclear weapons, took years of negotiation at leadership level—and one of the greatest geopolitical convulsions of the 20th century, the disappearance of the USSR—to attain.[34] Those reductions were not the product of a technical plan, except in the sense that they *did* require technical planning (how could they not have?). They, instead, followed a remarkable closing of the gap in trust between two long-time adversaries and a transformation in the international relations of the day. The TPNW's Article 4(2) 'notwithstanding' clause should not be mistaken for a provision that acknowledges this difficult reality of arms control. The TPNW reduces disarmament to arms elimination, and arms elimination to a ministerial function for an intergovernmental meeting. If disarmament were that easy, then we would have disarmed a long time ago. Yet any nuclear-weapon state, if it were in future to become party to the TPNW, would subject itself to this approach in legally binding terms.

TPNW Article 12 merits brief remark. Article 12 obliges each state party to 'encourage States not party to [the TPNW] to sign, ratify, accept, approve or accede to the Treaty, with the goal of universal adherence.' Article 12, obliging parties to evangelize the TPNW, resembles other recent treaty provisions that aim to bring about changes in international law at large.[35] Legal declaration, however, does not equate to geopolitical transformation, and it is the latter that disarmament requires.

Consequences of the TPNW and the missing negotiation piece

Though the TPNW is a treaty of limited subscription and articulates a prohibition that applies only to its parties,[36] the entry into force of the TPNW has potentially widespread consequences. Let us consider, first, how disarmament activists, disappointed with the NPT, met the TPNW with a groundswell of applause and denied that the new treaty will clash with the old; second, how the disarmament community's disappointment with the NPT and the realists' resiling from the NPT join to create a political vacuum that the TPNW is available to fill; and, third, how the TNPW, if it gains influence over the nuclear weapons discourse, will affect the US and its allies much more than it will our competitors.

NPT and TPNW hand in hand—or the absolutists' triumph?

The disarmament community marked with satisfaction the conclusion and, later, entry into force of the TPNW. Writers described the TPNW as 'a monumental achievement' and a 'tipping point' bringing to fruition 'widespread agreement' on disarmament;[37] a 'turning point,'[38] a 'major milestone in international law and nuclear security policy,'[39] 'a daring act of self-empowerment,'[40] and 'one of the most important diplomatic developments of the modern era.'[41] A triumphant tone entered disarmament discourse.

The disarmament community carefully assures countries whose accession to the TPNW they seek that the TPNW complements the NPT and, in any event, does not replace it.[42] A number of states agree. For example, states comprising the New Agenda Coalition (Brazil, Egypt, Ireland, Mexico, New Zealand, and South Africa) said before the 2020 NPT Review Conference that the TPNW 'complements and strengthens' the NPT, the earlier treaty remaining 'the cornerstone of the nuclear disarmament and non-proliferation regime.'[43] So the proposition that the TPNW does not displace the NPT, but functions *with* it hand in hand, has gained acceptance at governmental level.

Writing in 2020 in *Daedalus*, Harald Müller and Carmen Wunderlich posit that '[i]t is not ... preordained by the nature of the TPNW that it will damage the NPT.'[44] To say, as Müller and Wunderlich do, that the TPNW does not 'preordain […]' a damaging outcome offers scant reassurance, however. We should seek, instead, to estimate how *likely* a damaging outcome is. The divergence between the TPNW and the NPW suggests that the likelihood of a clash between the two is not low. Müller and Wunderlich venture the diffident proposition that a TPNW-instigated arms control disaster is not 'preordained' presumably for the very reason that they recognize that the likelihood of a clash is high.

Taking a somewhat firmer stance, Müller and Wunderlich say that '[t]he TPNW is no catastrophe to the NPT, but compatible with it,'[45] but here, too, they hedge. In their view, much turns on what parties and other actors do about the TPNW. According to Müller and Wunderlich: 'Whether the two treaties are compatible or not depends on how opponents and proponents of the TPNW handle their controversies. Reasonable policies can create a *modus vivendi*. Antagonistic policies can create incompatibility. Right now, the outcome is indeterminate' (172).

An 'indeterminate' outcome is no more reassuring than the 'not … preordained' outcome in which Müller and Wunderlich concede that the TPNW will 'damage' the NPT. The hedging, frustrating as it is, however, is not the only facet of Müller and Wunderlich's argument that merits critique. To say that whether or not two texts conflict depends on 'how opponents and proponents … handle their controversies,' is either to state the obvious: if parties do not join issue over the meaning of a text, then no issue will be joined—or it is to declare legal texts in general to be 'radically indeterminate.'[46]

If Müller and Wunderlich are stating the obvious, then they are stating a generality that does not apply in the face of the geopolitical facts. Whether policies are judged to be '[r]easonable policies' will differ, depending on which antagonist in the current competition between great powers is judging. Moreover, in a competitive setting, antagonists most likely *will* 'handle … controversies' in a manner that increases contestation. It is true—and a truism—that, if everyone lets a sleeping dog lie, then the dog will not bark. It is also true that contestants pursue strategies in which they do not wish every potential treaty conflict to remain silent.

If, instead, Müller and Wunderlich are not merely stating the obvious, then they espouse a view that legal texts—including the TPNW and NPT—are radically indeterminate—that is to say, too uncertain to allow confident conclusions as to their meaning. If *that* is what they mean, then these writers are difficult to address with ordinary methods of legal analysis.

There are contradictions in Müller and Wunderlich's account. They concede that '[t]he TPNW debate affects the normative order concerning nuclear weapons'[47] and, yet, deny that the debate will affect the NPT. Do they maintain that the NPT is not part of the 'normative order concerning nuclear weapons'? If so, then theirs is an incomplete definition of the normative order.

There is a further problem in Müller and Wunderlich's account of how the TPNW 'affects the normative order.' The normative order changes, and it is not only the *governments* of states that instigate it to change. In representative democracies, a diverse range of actors influence public policy. The TPNW is a rally point for disarmament activists. Disarmament activists exert influence on public policy in democracies. They do not, however, exert commensurate influence in societies under authoritarian rule, a problem to which I will return shortly. We should expect the TPNW to 'affect […] the normative order'—and, in particular, to affect the NPT.

I recalled in Chapter 1 that, in the years immediately after the ICJ's Advisory Opinion on *Nuclear Weapons*, the disarmament community took encouragement from the ICJ's interpretation of Article VI. Disarmament activists were hopeful that the Court's assertion that Article VI obliges disarmament as a *result*, not that states *endeavor* toward that result, had opened the door to rapid disarmament. The activists' hopefulness dwindled over time.[48] The ICJ's refusal to address the merits of recent claims by the Marshall Islands against the nuclear-weapon states under Article VI[49] probably brings that phase of disarmament aspiration to a close.

It is against the backdrop of their disappointment in Article VI that the disarmament community celebrated the arrival of the TPNW. Some commentators assert that the TPNW does not interfere with the NPT. However, policy makers would be mistaken to assume that without a fresh approach to arms control the NPT will continue to function in a TPNW age.

Non-proliferation abhors a vacuum

Policy makers responsible for international security and arms control might ask, if the NPT and its associated negotiating agenda were to fade, then so what? Surely, the important thing is that the US and its allies maintain a deterrent, not that we pursue arms control.

There are dangers, however, in a 'so what' response.

For one, the deterrent that we need is shaped by the security environment in which we operate. Non-proliferation shapes the environment. When non-proliferation *works*, it shapes the environment for the better. Disappearance of non-proliferation as an operative category of security policy does not enhance security in a world in which malign actors continue to seek nuclear weapons and those who already have them are emboldened to threaten to use them as an umbrella for conventional armed aggression. The first branch of this observation is straightforward: the decline of non-proliferation diplomacy hinders efforts to keep nuclear weapons out of the hands of malign actors (whether states or others). The second branch—observing that nuclear-weapon states (Russia and China) are conducting or threatening conventional armed aggression against non-nuclear-weapon states and that this situation, too, recommends maintaining a viable non-proliferation diplomacy—merits elaboration. Conventional armed aggression conducted by a nuclear-weapon state that uses the threat of nuclear escalation to dampen the defender's response (and that of the defender's supporters) is a potent incentive to proliferate, because it communicates to non-nuclear-weapon states that they are vulnerable to attack. In short, one reason to preserve non-proliferation diplomacy has not gone away—malign actors seeking to proliferate; and a new reason arises—non-nuclear-weapon states seeking the means to ensure their own self-defense.[50] In today's security environment, a viable NPT is more important than ever.

Indifference about the future of the NPT is unwise for a further reason. The disappearance of a realistic non-proliferation agenda would not leave a hole in this policy space. If the NPT were to fade, then this would create a vacuum, and non-proliferation abhors a vacuum. We now know what would fill it. The TPNW is the matter waiting to rush in. Evidence suggests that this is happening already.

Uptake of the TPNW in the first two years after it opened for signature was rapid.[51] At least some of the states parties to the TPNW appear to be taking seriously the obligation under Article 12 TPNW to 'encourage States not party to this Treaty to sign, ratify, accept, approve or accede to the Treaty, with the goal of universal adherence of all States to the Treaty.'[52] As I noted, by July 2024, the TPNW had 93 signatories, of which 70 were parties.[53] It is too early to say that the TPNW has overtaken the arms control and disarmament agenda; just as it is too early to say that the NPT has declined beyond recovery. However, the trends are not promising.

Pace Müller and Wunderlich, the NPT and TPNW are treaties in collision. The collision is between the NPT's negotiation provision and the TPNW's prohibitionary obligation. The NPT pledges its parties to negotiate toward arms control and disarmament; the TPNW obliges them to disarm. This is the difference between a treaty that nuclear-weapon states painstakingly crafted to address the world in which they existed—a world the lineaments of which are very much visible still today—and an aspirational exercise calling for radical change. The problem with the aspirational exercise is that it pledges countries to implement a disarmament end-state without regard to the risks that that end-state would entail if only *some* countries reached it; and without regard to the effort required if *all* countries with nuclear weapons are to agree to effective disarmament measures.

The 2022 NPR reiterates that the TPNW fails to serve the interests of international security. According to the 2022 NPR:

> While the United States actively pursues the goal of a world without nuclear weapons, it does not consider the Treaty on the Prohibition of Nuclear Weapons (TPNW) to be an effective means to reach that goal. The United States does not share the underlying assumption of the TPNW that the elimination of nuclear weapons can be achieved irrespective of the prevailing international security environment. Nor do we consider the TPNW to be an effective tool to resolve the underlying security conflicts that lead states to retain or seek nuclear weapons.[54]

This is a welcome statement. However, it does not go far enough.

I suggested in Chapter 1 that the 2022 NPR, focusing on the affirmative benefits of the NPT, errs in saying too little about the negative function of the NPT—the avoidance of worst outcomes. The NPR when it addresses the TPNW errs in a converse way. As expressed in the 2022 NPR, the US's concerns about the TPNW appear to be chiefly *negative*: the TPNW does not do what we need a non-proliferation policy to do. The deficiency in expressing our concerns in this negative way is that the TPNW's failings are not limited to what it does not do. The TPNW fails much more seriously in what it *does*: it gives the false impression that a declaration in legal form on its own will attain disarmament. In giving that false impression, the TPNW motivates civil society in a range of countries to pursue shifts in nuclear policy that not only will fail to attain disarmament but also will make disarmament harder than ever to attain.

Before turning to how we might re-invigorate NPT Article VI, and, by so doing, prevent a vacuum that the TPNW will fill, we need to remind ourselves of another important reason that we do so: the TPNW will influence *our* security strategy more than it will influence our competitors'.

The TPNW's asymmetric effect

The political and rhetorical effect of an arms control initiative differs from state to state. An authoritarian, one-party state that allows little room for public discourse on arms control and, in any event, is little influenced by public preferences, experiences the effect of an arms control initiative very differently from a representative democracy with an open society.[55] Difference in political or rhetorical effect is by no means the only asymmetry in arms control; asymmetries are pervasive here.[56] The asymmetry in how different states experience the TPNW's influence, however, is one that policy makers in the US and allied countries so far have done little to address. Our competitors are unlikely to ignore it. It gives further urgency to revitalizing the NPT.

Müller and Wunderlich posit that forbearance in how we interpret the NPT and the TPNW would allow the two treaties to coexist without tension. They insist that the new treaty is not a threat to the old. However, they nevertheless concede that '[a]t the heart of contestation [over the two treaties] are two security concepts: deterrence versus the immediate ban of nuclear arms, which result in fundamentally different ideas on how to pursue the road to "global zero".'[57] As between democratic and authoritarian states, the 'fundamentally different ideas' will have fundamentally different effects on policy. Authoritarian states will promote in democratic states 'security concepts' likely to shape the normative order in ways that the authoritarians favor. The US and its allies will exert little if any shaping influence the other way.

We have been here before. The uneven effects of disarmament rhetoric were visible in the 1980s. Jeffrey Herf, who studied the West German peace movement and its effects on Western alliance strategy, observed that the USSR 'pulled all available levers of Western German politics' in the months leading to the 1983 West German federal elections.[58] When George H.W. Bush, as US vice president, made an official visit to Berlin on February 1, 1983, a mass protest of anti-nuclear activists greeted him; no protests met Soviet officials on official visits to Bonn.[59] The US and its allies had to contend with 'an "antinuclear" movement that focused primarily on the nuclear weapons of the democracies.'[60] As Herf observed, the disarmament conversation of the 1980s was not novel in its one-sidedness; the campaign against the neutron bomb in the 1970s had displayed much the same asymmetry.[61] Another scholar of the Cold War, Matthew Evangelista, has noted that the USSR's KGB and East Germany's Stasi during the 1980s financed activists in Western Europe who campaigned for the withdrawal of US Pershing II and cruise missiles, while at home the USSR dispersed peace groups and interned their members in psychiatric hospitals.[62]

Even without an adversary acting behind the lines, as it were, to mobilize civil society in furtherance of its strategic aims, democratic states feel the effects of disarmament discourse more sharply than do others. William Van Cleave, who advised the Reagan Administration on nuclear arms control and, among other roles, served on the US Delegation to the Strategic Arms Limitation Talks, observed nearly 40 years ago that 'in the democratic states of the West there is *always* arms control, even without negotiated agreements. Arms are controlled and limited by the West's traditional values, by its political and budgeting process, and by the influence of the media and of public opinion.'[63]

We disregard this axiom—that, in our societies, there 'is *always* arms control'—at our peril. Its salience today has grown as the disarmament discourse has shifted from Article VI of the NPT to the absolutist prohibition of the TPNW.

It is all too easy to conclude that the TPNW, because its goals are unrealistic, will not affect security policy. The seemingly settled assumption that we *will* continue a strategically minded policy of nuclear deterrence, because disarmament is simply unworkable in present circumstances, lures too many in the policy community into complacence. A sound policy of nuclear deterrence is not inevitable. The TPNW exerts an influence against realistic arms control and toward an absolutist approach to disarmament, yet policy makers have done little to reinvigorate the NPT, the treaty that offers the best available answer to the TPNW.

Enticing some policy makers to learn to stop worrying and ignore the TPNW, the five nuclear-weapon states are unanimous in their public opposition to the TPNW. According to the P5 Joint Statement on the TPNW (October 24, 2018), 'The TPNW will not be binding on our countries, and we do not accept any claim that it contributes to the development of customary international law; nor does it set any new standards or norms.'[64] China, in addition to joining the P5 Joint Statement, separately 'emphasized that the [NPT] is the cornerstone of the international nuclear non-proliferation system and an important component of the international security architecture' and that China joins the unanimous rejection of the TPNW.[65] According to China, 'China respects the role of the Non-Proliferation Treaty as the cornerstone of the international nuclear non-proliferation regime.'[66] But civil society in China (and in Russia) has little or no influence over the nuclear weapons strategy that those countries pursue. The effect of the TPNW is asymmetric, even as the five nuclear-weapon states join to express a shared interest in tempering the absolutism behind it. A degree of caution is in order, therefore, when we consider the stated opposition of Russia and China to the TPNW. Even if they are sincere, these countries are not affected the same way as ours by the suasive influence of disarmament discourse.

The US in particular needs to consider how disarmament discourse affects like-minded countries, including the closest US allies. If the TPNW comes to occupy the political space that heretofore belonged to the NPT, then serious challenges will arise for US security diplomacy. Public affirmations by allies and like-minded countries that it is still the NPT that matters, though no doubt sincere, should no more comfort US leaders than affirmations to the same effect by China or Russia. For example, countries comprising the Vienna Group of Ten (Australia, Austria, Canada, Denmark, Finland, Hungary, Ireland, the Netherlands, New Zealand, Norway, and Sweden)[67] stated in the Preparatory Committee leading to the NPT 2020 Review Conference that it is still the NPT that matters. They reiterated their 'full commitment' to the NPT, which they affirmed is '*the* cornerstone of the international nuclear disarmament and non-proliferation regime' (emphasis added).[68] But public opinion shifts, and, in democracies, when it shifts, it has the potential to carry government policy in its train.

A shift of opinion in the free world toward the TPNW is no mere conjecture. Australia, under a coalition government comprising the Liberal Party and National Party from August 2018 through May 2022, voted repeatedly against the TPNW in the UN General Assembly.[69] The coalition government, after the Australian federal election of May 21, 2022, was replaced by a Labor Party government. Australia, on October 28, 2022, changed its approach: instead of casting a negative vote against the TPNW in the General Assembly, Australia abstained.[70] Disarmament groups noted the change from nay vote to abstention with approval.[71]

Even in the UK, one of the two nuclear-weapon states that are US allies, questions arose before the July 4, 2024 General Election as to the Labour Party's commitment to nuclear deterrence. The Labour Party's 2024 Manifesto stated that the Party's 'commitment to the UK's nuclear deterrent is absolute.'[72] However, a number of MPs who voted in 2016 against renewing the Trident submarine program, basis of the UK's nuclear deterrent, have served on the Labour Party's leadership team[73] and more sit in the back benches. The experience of the several Conservative prime ministers immediately before Labour Prime Minister Sir Keir Starmer suggests that it is not easy for a UK Cabinet to ignore an activist backbench. Trident, not the TPNW, was the focal point of debate over nuclear deterrence in the 2024 UK General Election, and though not all opponents of Trident are supporters of the TPNW, disarmament activism has no small purchase in UK politics.[74]

There are also occasional calls in Western countries for a divestment campaign to target companies in the nuclear supply chain.[75] Farther afield, the TPNW continues to receive support among states in the Non-Aligned Movement.[76]

If the TPNW gains further adherents, then it might well offset, even negate, diplomatic advantages that are emerging from other trends. For example, commentators have suggested that China's failure to condemn Russia for the latter's war of aggression against Ukraine has eroded willingness among governments in Eastern and Central Europe to transact with China.[77] While no doubt Russia's aggression will give pause to countries in that region before they adopt a prohibitionist stance against nuclear weapons—consider, for example, the readiness of Finland and Poland to host nuclear weapons[78]—the TPNW, if it continues to attract support in civil society, pushes the other way. No symmetric effect arises from the TPNW in China or Russia.

THREE

Article VI Interpreted and Applied

At first appearance, NPT Article VI presents no special problem of interpretation.[1] Article VI is a negotiations provision. Many treaties contain provisions of that kind. However, as we saw in Chapter 1, contestation has arisen over Article VI. On the better reading, Article VI is a self-contained provision, not a bargaining chip for non-nuclear-weapon states to challenge the non-proliferation regime that forms the heart of the treaty. Confusion arises, when parties and commentators describe the NPT as an ongoing bargaining process in which parties compete over how the so-called 'three pillars' apply. Much of the confusion owes to the thesis that, because negotiating parties in the 1960s arrived at these provisions through a *quid pro quo*, the non-proliferation provisions apply only if the nuclear-weapon states implement Article VI, the negotiation provision, in a manner that non-nuclear-weapon states approve. Most treaties reflect bargains reached when they were drafted. But a drafting history that suggests that a given treaty term was the result of a bargain does not make that term an invitation to further bargaining today. To discern the meaning of a treaty, one looks first to the terms that the parties adopted, not to the process that led to adoption. Articles II and III, as written and adopted, do not express themselves as conditioned upon Article VI or on any other provision of the treaty.[2]

Even if we put aside the conditional reading of Articles II and III, questions still remain as to the meaning of Article VI itself.

For one, a question was raised at the end of the Cold War about the continued scope of application of Article VI:

considering that the superpower 'arms race' that existed when the NPT was drafted ceased in the early 1990s, does the 'arms race' limb of Article VI still apply? Statements by US officials in diplomatic settings have been read to suggest that it might not. The 'arms race' limb has renewed salience, however, in light of the arms buildup on which China has embarked.

A further question has puzzled policy makers practically since the NPT was concluded: what, precisely, does the 'pursue negotiations' clause of Article VI oblige a state party to do? This question, too, I will argue (in Chapter 4) China's buildup once more heightens as a policy-making concern.

Let us briefly consider the first question—whether the 'arms race' limb of Article VI continues to apply—before we take a closer look at the 'pursue negotiations' clause.

The text and a post-Cold War question

NPT Article VI, to recall, reads as follows: 'Each of the Parties to the Treaty undertakes to pursue negotiations in good faith on effective measures relating to cessation of the nuclear arms race at an early date and to nuclear disarmament, and on a treaty on general and complete disarmament under strict and effective international control.'[3]

In some sense, though not necessarily in the legal sense, US officials not long after the end of the Cold War suggested that the 'nuclear arms race' referred to in Article VI was exclusively that between the US and USSR.[4] We should take care, however, before we conclude that Article VI is limited to a particular arms race. On the better view, the 'arms race' limb continues to apply.[5] Let us consider briefly the reasons for the continued application of that part of Article VI.

To begin with, nothing inherent in the expression 'arms race' excludes applying the arms race limb of Article VI in the current geopolitical setting. Weapons modernization programs of the US, the UK, and France have not approached the scope of China's buildup, but nothing in the expression 'arms race' necessarily entails a symmetric relationship of action and reaction between nuclear rivals. Recent scholarship has suggested that the relationship between the nuclear weapons programs of the US and USSR, in the most febrile days of the Cold War 'arms race', was not very close: each superpower pursued its nuclear strategy with regard to a range of considerations, not only, or even mainly, in response to the other's nuclear weapons capability.[6] Even at the time, informed observers doubted whether Soviet nuclear behavior was in every instance a reaction to steps that the US had taken. Harold Brown, who served as US Secretary of Defense from 1977 to 1981, put it succinctly: 'Soviet spending [on nuclear arms] has shown no response to U.S. restraint—when we build, they build; when we cut, they build.'[7] However, nobody thought at that time that the 'arms race' limb did not apply. The magnitude and character of the 'arms race' change over time, as the NPT's drafters would have recognized, and nothing in the text makes the 'arms race' limb hinge on either.

What of the suggestions by US officials that the end of the Cold War resulted in the lapsing of *the* arms race and thus of the 'arms race' limb?

The end of the Cold War was a momentous event. However, the NPT does not condition its continued effectiveness upon the continuation of a particular geopolitical contest. Article VI refers to '*the* arms race,' but there is no limiting term in the NPT to stop states parties applying Article VI to an arms race outside the historical US–USSR relationship.

In the text of the NPT, we find further grounds to treat the 'arms race' limb of Article VI as a provision that still has legal effect. Article VI addresses 'the Parties to the Treaty.' It is not a provision limited to any one subgroup of parties. Moreover, it addresses the parties individually—'each' of them—not a compound unit of parties. Article VI obliges each of the parties to 'pursue negotiations in good faith.' Even though the US–USSR bilateral relationship was on the minds of policy makers at the time, a reading that limits Article VI to that particular bilateral relationship does not accord with the complete scope of addressees explicit in the text.

Also of some interpretative significance, when we evaluate the contention that the 'arms race' limb applied only to the US–USSR arms race, and, thus, lapsed when the arms race between those two states ended, no provision of the NPT distinguishes among different *nuclear-weapon states*. The NPT, famously, differentiates between states that had manufactured or exploded a nuclear weapon or other nuclear explosive device prior to January 1, 1967 and those that had not—that is to say, between nuclear-weapon parties and non-nuclear-weapon parties. There is no category in the NPT of an 'arms-race Party' or a 'non-arms-race Party.'[8] Nor is there a time-limit on 'arms race'—such as the time-limit distinguishing between nuclear-weapon states and others. Among the nuclear-weapon parties, the treaty makes no distinction.

Confusion about the 'arms race' limb arises chiefly from the second Bush Administration.' Ronald Bettauer, State Department Deputy Legal Adviser, at a 2006 bar association event said that 'the nuclear arms race between the United States and Russia has in fact ended.'[9] Ambassador Sanders at the 2005 NPT Review Conference said that 'the nuclear arms race referred to in Article VI is over.'[10] The conclusion of the arms race to which these officials referred was an observation of fact. An observation of fact is not in itself an expression of a change in legal policy of a state. Moreover, the burden for establishing a change of *international* policy of a state is high; statements by a legal advisor or an ambassador, even a high-ranking one, are not convincing evidence of a shift as significant as that which would have been entailed if the US had intended to declare that it no longer viewed the Article VI 'arms race' limb as applicable law. The statements do not demonstrate a change of US understanding of Article VI.[11] Moreover, the NPT is not a unilateral act; it is a multilateral treaty. Considerably more than a change of *US* understanding would have been needed to change Article VI.[12]

Chapter 4 will consider China's nuclear weapons buildup and will suggest an argument for invoking Article VI, including its arms race limb, in US and allied diplomacy in response to that strategic challenge. As I have briefly argued here, Article VI, including its arms race limb, applies as much today as it had to the earlier arms race that had animated states to draft and adopt the NPT.

But to say that Article VI continues to apply in full is not to say what, precisely, Article VI obliges a state to do. To arrive at an understanding of what Article VI obliges, let us consider requirements to negotiate under international law and the particular terms of Article VI.

Negotiation requirements generally

The negotiation clauses that have received the fullest ventilation under international law are those requiring that parties negotiate as an antecedent to some (usually compulsory and binding) dispute settlement mechanism.[13] For example, parties may have recourse to the dispute settlement mechanisms under Part XV of the UN Convention on the Law of the Sea (UNCLOS) after they have attempted to negotiate a settlement of their dispute.[14] Considering UNCLOS Article 83, paragraph 1 (which addresses delimitation of the continental shelf), the International Court of Justice (ICJ) in *Maritime Delimitation in the Indian Ocean* observed that the provision contains a requirement that 'there be negotiations conducted in good faith, but not that they should be successful.'[15] A negotiation requirement of that kind is an obligation of 'best efforts' (sometimes referred to as an obligation of conduct); it is not an obligation to achieve a specified result.

Indeed, jurists have underscored that a negotiation prescribed in such terms does not necessarily achieve a result at all:

> [E]ven where there is an obligation to negotiate, negotiations do not constitute, as such, a method of

> dispute settlement because they may or may not lead to a settlement, depending wholly or partly on the position of one of the States concerned. If States agree to negotiate but leave all their options open as to the outcome of those negotiations, they have not necessarily agreed to a method of *settlement*: it is equally possible that the dispute will not be settled.[16]

That an obligation to negotiate might not lead to a settlement does not mean that the obligation is without legal consequence. Obligations of this type often stipulate negotiation as a requirement that parties must fulfill before invoking a judicial or arbitral dispute settlement clause. Where an obligation of this type applies, a party that does not wish to litigate or arbitrate is likely to argue that its adversary has failed to negotiate.[17] Where a party has argued to that effect, adjudicators have had to consider whether particular conduct—that is, conduct by the party that wishes to litigate or arbitrate—constitutes an attempt to negotiate.

Courts and tribunals concur that paying mere lip service to an obligation to negotiate does not suffice. The obligation requires that the state concerned 'make [...] a *genuine attempt*.'[18] For an attempt to be genuine, the parties must have 'conducted themselves in such a way that negotiations may be *meaningful*.'[19]

There are many ways to conduct a meaningful negotiation. However, not every diplomatic contact or exchange is a negotiation. As the ICJ observed in *Georgia v. Russia*, 'negotiations are distinct from mere protests or disputations. Negotiations entail more than the plain opposition of legal views or interests between two parties, or the existence of a series of accusations and rebuttals, or even the exchange of claims and directly opposed counter-claims.'[20]

The negotiation requirement under consideration in *Georgia v. Russia* was a pre-condition for recourse to the ICJ. Though, as noted, this is the kind of negotiation requirement that courts most often have had occasion to apply, not all obligations to negotiate are a pre-condition for recourse to a court. The obligation in Article 5, paragraph 1, of the North Macedonia[21]–Greece Interim Accord was not a pre-condition of that kind. It was an obligation to negotiate 'with a view to reaching agreement on the difference [over the name of North Macedonia].'[22] The obligation, in other words, was a free-standing obligation, not a prerequisite to judicial or arbitral procedure.[23] The ICJ concluded that the parties had fulfilled the obligation. The name of North Macedonia remained in question at that time, but the obligation was that the parties *try* to settle the question of the name, not that they succeed in settling it. In reaching its conclusion, the Court noted that North Macedonia 'showed a degree of openness to proposals that differed' from its own preferred outcome; and Greece, too, 'changed its initial position.'[24] The factors that the ICJ considered in order to judge whether North Macedonia and Greece had negotiated to seek a settlement of their difference are much the same as those that the Court and adjudicators elsewhere have considered in order to judge whether parties have satisfied requirements to negotiate contained in third-party dispute settlement clauses.

These recent cases reflect an understanding of negotiation requirements that has prevailed for some time. In the *North Sea Continental Shelf Cases*, which Germany, the Netherlands, and Denmark argued when the NPT was in its final drafting stages,[25] the ICJ said that negotiation has not taken place where either party 'insists upon its own position without contemplating any modification of it.'[26] Then, as now, mere repetition by one state of its preferred position will not satisfy an obligation to negotiate. When we turn in Chapter 5 to consider possible topics of nuclear arms control that the US might negotiate with China, we should keep in mind that parties must show flexibility if they are to satisfy an obligation to negotiate.

Jurists also agree that, when parties commit in legally binding terms to negotiate, they commit to negotiate in good faith.[27] 'The principle of good faith is . . . one of the basic principles governing the creation and performance of legal obligations.'[28]

International law supplies few precise signposts for distinguishing between good faith and its absence. Judging the inner state of mind of a state is an exercise fraught with difficulty,[29] and, so, the focus here has been on outward signs.[30] Where parties have successfully resolved a dispute through negotiation, one might infer that the parties negotiated in good faith. However, the continuation of a dispute says nothing in itself conclusive as to the parties' good faith. Because a negotiation might lead to one of any number of possible resolutions of the parties' dispute, or to no resolution at all, the outcome of a negotiation does not say much, if anything, as to whether the parties conducted themselves in good faith in the course of negotiation. 'Two parties, each acting in good faith, or not demonstrably in bad faith, can fail to reach agreement.'[31]

Where evidence shows that a party has acted *in bad faith*, the party clearly has failed to meet its obligation. 'Negotiation in bad faith' is a contradiction in terms. The principle of good faith is not a separate source of obligation where negotiations are concerned;[32] observance of good faith is integral and necessary to negotiation.

Where a party, though obliged to pursue negotiations, has refused to, that too presents a clear case. In that case, the matter is one of complete non-performance, rather than a defect vitiating an attempted negotiation.[33]

A negotiation requirement is a requirement to negotiate about a particular thing. A party does not fulfill a negotiation requirement by negotiating about *X* when the requirement is to negotiate about *Y*. That is to say, 'the subject-matter of the negotiations must relate to the subject-matter of the dispute.'[34] The ICJ refused to adjudicate Georgia's claims that Russia had violated the Convention on the Elimination of All Forms of Racial Discrimination (CERD) even though Georgia had negotiated with Russia; the problem was that Georgia had not negotiated with Russia about CERD subject matter.[35]

A treaty might express only at a high level of abstraction the subject matter that it obliges its parties to negotiate about, and the subject matter might concern an overall situation or problem, rather than a classic bilateral dispute such as those litigated from time to time in court. However, even where the texture of a negotiation requirement is relatively open, the requirement still addresses a particular subject matter.

Finally, a negotiation requirement entails that each party observe a certain restraint in its conduct concerning the subject matter. Under the principle of good faith, which as we saw inheres in the negotiation requirement, each party must refrain from conduct that creates a *fait accompli* or that aggravates the parties' difference. Chapter 4 will return to this point about restraint, which is consequential for nuclear weapons policy.

The negotiation requirement in Article VI

Article VI of the NPT obliges the parties 'to pursue negotiations in good faith on effective measures relating to cessation of the nuclear arms race at an early date and to nuclear disarmament.'[36] Article VI relates to the aspirational statement in the Preamble '[d]eclaring [the parties'] intention to achieve at the earliest possible date the cessation of the nuclear arms race and to undertake effective measures in the direction of nuclear disarmament.' The preambular language underscores that the parties drafted Article VI to place them in the *direction* of disarmament through practical, negotiated steps. Article VI is not a prohibition of nuclear weapons. Article VI as drafted and adopted is, instead, a negotiation clause, intended to move the parties *in the direction* of arms control and eventual disarmament.[37]

One needs more than one party to negotiate. As the 2022 NPR observes, '[n]egotiation requires a willing partner operating in good faith.'[38] The refusal of one party to negotiate, however, does not constitute a violation of the Article VI obligation by another party. It would make no sense in this setting if the nonfeasance of one were to result in the attachment of international legal responsibility (that is, liability) to the other.

Conversely, the refusal of one party to negotiate does not absolve the other of its own failure to pursue negotiations. At least up to a point: the other party is not duty-bound to keep in perpetual motion toward a negotiation that the refusing party's intransigence assures will never start.

Much as one needs at least two parties to have a negotiation, one also must have at least one topic to negotiate. As I recalled earlier, an obligation to negotiate is an obligation to negotiate about a specified subject matter. Article VI does not specify the subject matter for negotiations in detail, but, instead, provides a description in open terms under two heads—first, 'effective measures' and, second, 'a treaty.' Both these heads are particularized, to a degree, in subordinate clauses. 'Effective measures' are to relate to two matters, these being identified in a two-limbed subordinate clause: '*cessation of the nuclear arms race at an early date*' and '*nuclear disarmament.*' As for the treaty, this is to be 'on general and complete disarmament under strict and effective international control.' To call for such a treaty is ambitious. The treaty head of Article VI covers not just nuclear arms, but all arms ('*complete* disarmament'). The US position has been that 'the language contains no suggestion that nuclear disarmament is to be achieved before general and complete disarmament.'[39] This position accords with an understanding that disarmament must take place through a process that enhances, rather than jeopardizes, geopolitical stability.

The 'effective measures' head,[40] by contrast, encompasses only nuclear arms—the nuclear arms race and nuclear disarmament being each the subject of one of the two limbs under the 'effective measures' head.

In light of their centrality to Article VI, the 'pursue negotiations' clause and the 'effective measures' clause each merits further remark.

The 'pursue negotiations' clause

Though directional, and though the parties drafted it conscious that an obligation of immediate disarmament is impractical, Article VI stipulates a binding legal obligation. As I discussed earlier in this chapter, like other negotiations provisions, Article VI binds the parties to negotiate. It is worth drawing attention to the binding legal obligation in Article VI, because not all commentary on Article VI accepts that it has this obligatory content, and, perhaps, nor do all states parties.

China, which, as we will see in Chapter 4 is of central concern when it comes to arms control today, does not appear to have given a complete statement of its understanding of Article VI. One Chinese legal writer said in 2009 that Article VI is merely 'hortatory or inspirational in nature.'[41] Reducing Article VI that way is difficult to accept, for the terms of Article VI are mandatory ('[e]ach of the Parties . . . undertakes . . .').

At least one US policy thinker with high-level experience in arms control negotiations, Christopher A. Ford, once suggested that the words 'to pursue' have a limitive effect on the negotiations clause.[42] The effect of the words, in Ford's view at the time, is not to reduce the mandatory terms of Article VI to a merely 'hortatory or inspirational' message, but it *is* to confine the obligation to exploratory steps or overtures and thus to distinguish it from an obligation to negotiate as such. In that view, an obligation to *pursue* negotiations is something less than an obligation to *negotiate*; and, thus, Article VI is not a negotiation provision but, rather, a pre-negotiation procedural clause stipulating that the parties will investigate the *possibility* of negotiating.[43]

Is this interpretation convincing—that is, does the phrase 'to pursue' in the negotiations clause of Article VI qualify the clause so that it requires less than actual negotiations?

On at least three grounds, it is arguable that the 'to pursue negotiations' clause does not differ from a clause requiring actual negotiations.

First, the ICJ and individual countries have maintained, with some consistency, that Article VI expresses a legal requirement to negotiate. Second, a negotiation requirement, whether expressed with the verb 'to pursue' or without it, embraces the pursuit and, so, any difference, to this extent, is one of form, not function. And third, the French-language version of the NPT and examples from other treaties in settings where courts use both French and English as authentic languages illustrate that 'to pursue negotiation' is synonymous with 'to negotiate' when used in settings such as Article VI. Each of these points merits elaboration.

Turning first to the ICJ, the Court in the *Nuclear Weapons* Advisory Opinion referred to 'the full importance of the recognition by Article VI of the [NPT] of an *obligation to negotiate* in good faith a nuclear disarmament.'[44] The ICJ in the three cases on *Negotiations Relating to Cessation of the Nuclear Arms Race and to Nuclear Disarmament* quoted that statement from the Advisory Opinion with approval.[45] The ICJ reads Article VI to require that the parties negotiate. Nothing in the Court's view identifies the Article VI 'obligation to negotiate' as allowing the parties to remain forever a step removed from negotiation. The Court, instead, views it as a commitment to do what it says. The faults in the 1996 Advisory Opinion notwithstanding, the ICJ's view of Article VI, to this extent, accords with the plain reading of the text.

There is much in the ICJ's nuclear weapons advisory opinion to question; the magisterial dissent of Vice-President Schwebel identified the main difficulties.[46] However, the Court's understanding that the 'to pursue negotiations' clause requires actual negotiations is well supported by a range of international practice. Over a century of international practice in various settings evinces a shared understanding that to pursue negotiations means to enter into and conduct negotiations, not just to deliberate about negotiating.

Australia, for example, in its Declaration made upon signature of the NPT, 'welcome[d] the call in Article VI . . . for negotiations.'[47] Germany, in its understandings expressed upon signature, said that 'the Parties to the Treaty will commence without delay the negotiations on disarmament envisaged under the Treaty, especially with regard to nuclear weapons.'[48] Australia and Germany understood Article VI to concern negotiations, not an antecedent exploratory phase distinct from negotiations.

The Permanent Court of International Justice, precursor to the ICJ, some 90 years ago used 'to pursue [negotiations]' to mean 'to negotiate.' In its 1931 Advisory Opinion on *Railway Traffic between Lithuania and Poland*, the Permanent Court interpreted a Resolution of the Council of the League of Nations that placed a legal duty on two states to negotiate: 'The Court is indeed justified in considering that the engagement incumbent on the two Governments in conformity with the Council's Resolution is not only to enter into negotiations, but also to pursue them as far as possible, with a view to concluding agreements.'[49]

In the opinion of the Permanent Court, which the ICJ would echo 65 years later, 'to pursue' meant to conduct negotiations, which entails the parties having entered into them. Pursuit, in other words, is not something that parties do separately from negotiating.

A treaty drafter has words to denote a procedural antecedent or other prior step distinct from negotiation. Consider again the UN Convention on the Law of the Sea; in particular, that treaty's Article 283. Under UNCLOS Article 283, which is titled 'Obligation to exchange views,' parties to a dispute 'shall proceed expeditiously to an exchange of views regarding its settlement by negotiation or other peaceful means.'[50] As drafted, this provision is clear that parties have forms of diplomatic contact that are not the same as negotiation. Under Article 283, parties, first, exchange views; they sometimes then agree to negotiate; they sometimes instead agree to employ 'other peaceful means.' UNCLOS Article 283, in requiring an exchange of views, indicates something that NPT Article VI does not: a diplomatic step prior to negotiation and necessarily separate and distinct from it.[51]

Finally, it is to be observed that the dictionary definition of 'to pursue' in the English language, and the use of the term in at least one of the other authentic languages of the

NPT, support the same interpretation. The phrase 'to pursue' in English may mean to carry on or proceed with (as in an action or procedure); this is one of some 11 variants and sub-variants indicated in the *Oxford English Dictionary*.[52] However, it is the chief variant concerning legal procedure. Reports of decided cases contain many instances in which courts in English-speaking countries have used the expression 'pursue negotiation' in accordance with the dictionary.[53] They use the expression to mean the action of continuing a negotiation that already has begun.

In the French language, NPT Article VI states that '*[c]hacune des Parties au Traité s'engage à poursuivre de bonne foi des négociations*.' The French verb *poursuivre*, like the English verb 'to pursue,' connotes the continuance of a procedure under way. The connotation is discernible from judicial practice in bilingual courts. Thus, for example, in a judgment of the European Court of Human Rights (where, like in the NPT, the English and French texts are equally authentic),[54] <<*le refus par un requérant d'entamer ou de poursuivre des négociations avec les authorités d'un Etat défendeur*>> is equivalent to 'an applicant's refusal either to enter into or continue negotiations with the authorities of a respondent State.'[55] Both the contrast between <<*entamer*>> and <<*poursuivre,*>> and the equivalency between <<*poursuivre*>> and 'continue,' suggest the proper interpretation of the French verb in NPT Article VI: <<*poursuivre*>> as used in Article VI means that the parties are obliged to negotiate. The Court of Justice of the European Union, which, like the Court of Human Rights, publishes texts of its judgments in English and in French,[56] deals with 'to pursue' much the same way. Thus, <<*la Commission souhaitait poursuivre à son compte les négociations*>> is in its English form 'the Commission wished to continue the negotiations on its own behalf.'[57] As far as I am aware, no other variant of <<*poursuivre*>> appears in the jurisprudence of either of these treaty-based courts.

It goes without saying that one needs more than one party to negotiate; and the parties need to take steps to set negotiations in motion. An obligation to pursue negotiations necessarily entails preliminaries if parties are to fulfill it. The practical realities of negotiation are reflected in the interpretation and application of negotiation clauses. An obligation to negotiate such as that under NPT Article VI does not come with a pre-fixed schedule or demand a constant state of engagement without preamble or pause.[58] As I noted earlier recalling the jurisprudence, an engagement, to be a negotiation, must be *meaningful*. This element of negotiation itself suggests that satisfying a negotiation requirement involves outreach, exploration, and a rhythm adaptable to an evolving substantive colloquy. Negotiation allows that the parties are not constantly at the table.[59]

Moreover, because it takes more than one party to have a negotiation, negotiation entails invitation, agenda-setting, and, often, trial-and-error.[60] It does not constitute one party's breach of a negotiations clause that another party is intransigent. If breach happened that way, then the negotiations clauses contained in many dispute settlement provisions would present a moral hazard: the party that wishes to avoid the arbitration tribunal or court merely would have to drag its feet, and, refusing to negotiate, would deprive the other party of the chance to fulfill the negotiation requirement. That is not how courts or tribunals interpret and apply negotiation clauses.

In the obligation to negotiate inheres the pursuit of negotiation. NPT Article VI makes this relationship explicit, which is, arguably, better drafting; but treaties that imply the relationship have not led courts and tribunals to interpret the obligation to negotiate differently.[61]

The 'effective measures' clause

Academic writers have lamented that the 'effective measures' clause of NPT Article VI leaves a wide discretion in each state party.[62] For reasons that I will set out briefly, the 'effective measures' clause does leave a wide discretion in each state party, but the discretion is within the limits of the legal duty that Article VI imposes.

The parties to whom the first sentence of Article VI is addressed—'[e]ach of the Parties to the Treaty'—have accepted an obligation: they have 'undertake[n]' to do something. They have undertaken 'to pursue negotiations in good faith.' There is no point in negotiating without a particular object, and, so, the predicate of the sentence is comprised of that active part ('to pursue negotiations ...') and a pair of subordinate clauses, these giving the negotiations two particular objects. As I noted earlier, the first subordinate clause is '*on effective measures relating to cessation of the nuclear arms race at an early date and to nuclear disarmament.*' The second is '*on a treaty on general and complete disarmament under strict and effective international control.*' Here, then, are the objects toward which the states parties are to negotiate.

The core element of Article VI is the obligation to negotiate. It follows that the parties have not yet reached agreement settling the differences between them in respect of each object toward which they are to negotiate: if they had settled their differences in that respect, then it would be to no purpose to oblige them to negotiate toward those objects. Taking the second object of negotiation—a treaty—parties negotiating toward a treaty will have differences of view as to what the treaty should say. Parties negotiating toward a treaty addressing a matter of the complexity and gravity of 'complete disarmament . . .' will have particularly diverse and contentious views as to what *that* treaty should say.[63] The parties also will differ, when they negotiate toward the other object, 'effective measures relating to cessation of the nuclear arms race ... .' The object of a negotiation, whether prosaic or fraught with geopolitical concerns, is, by definition, an object of difference.

When parties come to the table to negotiate, they likely will have different conceptions of how to achieve the object toward which they are to negotiate. The discretion in Article VI as to 'effective measures' accommodates the parties' different conceptions. Accommodation here, however, must not be interpreted so far that it excludes any possibility of a successful negotiated result. The parties will have different conceptions of 'effective measures,' but that they differ is not in itself ground for refusing to negotiate.

Even a negotiation toward a less complex object than nuclear arms control, such as settling Greece and North Macedonia's difference as to the latter's name, entails back-and-forth and takes considerable time. The complexity of arms control being so much greater than many other objects of negotiation, fashioning 'effective measures' will take a great deal of effort, even if political differences, such as those that made the Greece–North Macedonia negotiations difficult, are set aside.[64]

To the extent that each NPT party has discretion to interpret the 'effective measures' clause, interpretations must remain within the bounds of Article VI as a whole. It is unconvincing to read a subordinate clause of a negotiation provision in a way that eliminates the obligation to negotiate.[65] A natural reading of Article VI is that a purpose in pursuing negotiations is to work through permutations of 'effective measures' that states are likely to conceive in furtherance of arms control. Differences between states as to what measures might be 'effective measures' do not efface the obligatory direction in Article VI to negotiate. To the contrary, it is because states have differences in regard to arms control that negotiation is the approach that they have accepted and that Article VI requires.

FOUR

China and the NPT

Whether, as in the Biden Administration's National Security Strategy, one describes China as the 'pacing challenge,'[1] or, as in the Trump Administration's National Security Strategy, one identifies China together with Russia as one of a pair of 'revisionist powers'[2] that 'are contesting our geopolitical advantages and trying to change the international order in their favor,'[3] China affects the strategic environment in fundamental ways.[4] China has made clear that it aims to compete with the US and its allies across multiple domains.[5] Perhaps most striking is China's nuclear weapons buildup. It remains unclear precisely what strategic purpose China pursues through its buildup, but the quality and quantity suggest that China's intentions are transformative.[6] Whatever its strategic purpose, the buildup has implications for the NPT. In view of the shift of disarmament discourse to the nuclear ban treaty, the TPNW, and the difficulties that the fading of the NPT would entail, it is under the NPT that the US and its allies should craft a diplomatic strategy to enhance our overall response to China—and to pursue the best chance of reviving nuclear arms control.

'Breathtaking expansion' and missing response

The 2022 Nuclear Posture Review acknowledges the stark reality of China's nuclear weapons buildup. Using language similar to that in the 2022 National Security Strategy, the NPR refers to China as 'the overall pacing challenge for U.S. defense planning and a growing factor in evaluating our nuclear deterrent.'[7] The NPR repeats the publicly available estimate that China pursues a nuclear arsenal of 'at least 1,000 deliverable warheads,' with 2030 as the target date for attaining that new quantity of armament.[8] For its part, the Pentagon's annual report for 2022 on China—*Military and Security Developments Involving the People's Republic of China*—indicates that by 2035 China will have 1,500 warheads.[9] The NPR says that it is likely that China will use the envisaged arsenal as 'leverage … for coercive purposes' in pursuit of expansionist aims.[10] As President Trump had said two years before, 'China is surging.'[11]

Shortly after the public release of the 2022 NPR, US Navy Admiral Charles A. Richard, commander of US Strategic Command, warned that '[a]s I assess our level of deterrence against China, the ship is slowly sinking'; China is 'putting capability in the field faster than we are.'[12] The impact on strategic stability is patent. Before that, Admiral Richard, in testimony before the House Armed Services Committee, observed that China

> continues the breathtaking expansion of its strategic and nuclear forces with opaque intentions as to their use … The strategic security environment is now a three-party nuclear-peer reality, where the PRC and Russia are stressing and undermining international law, rules-based order, and norms in every domain. Never before has this Nation simultaneously faced two nuclear-capable near-peers, who must be deterred differently.[13]

There is no denying that China is embarked on a nuclear weapons buildup of considerable scope, and we must entertain that China's purpose in the buildup is fundamentally to alter the strategic environment.[14] However, the US and its allies have not yet met China's 'breathtaking expansion' with a fresh approach to arms control diplomacy.

Diplomacy is one piece of a larger strategic response. Many experts agree that the US and its allies need to invest in deterrent capabilities as well. For example, Ryan Tully, noting that US withdrawal from the Intermediate-Range Nuclear Forces (INF) Treaty[15] in 2018 by the Trump Administration[16] opens the door for the US to develop and deploy new defensive systems, counsels accelerated deployment of intermediate-range missiles in the Indo-Pacific.[17] Keith B. Payne and Matthew R. Costlow call for enhancement across the full 'spectrum of regional and strategic nuclear forces' in order to deter China.[18] Legislators have as well. For example, US Senator Jim Inhofe and US Representative Mike Rogers criticized scaling back of nuclear-weapon programs that the Trump Administration had supported.[19] In the UK, the Integrated Review in March 2021 reversed the decision, that had been taken in 2010, to reduce the UK's nuclear warhead stockpile and, instead, announced a slight increase.[20]

The capabilities that suffice for deterrence depend, in part, on the capabilities that a challenger holds. As Schelling defined it, '[d]eterrence … is concerned with influencing the choices that another party will make, and doing it by influencing his expectations of how we will behave.'[21] Our challengers' capabilities limit how we *can* behave and, *a fortiori*, limit our challengers' expectations of how we *will* behave: a challenger will confidently expect that we will not do that which we *cannot* do, assuming that the challenger knows that we cannot do it. Arms control in the past placed effective, verifiable limits on the capabilities of both parties.

It is hard to see much chance for arms control initiatives today, so long as the nuclear-weapon state with the most ambitious plans for enlarging its arsenal does not join them.

What does the NPT have to say about China's nuclear weapons buildup? And how might we use the NPT to place effective pressure on China to come to the negotiating table?

What the NPT does not do—and what it does

NPT Article VI is not a provision prohibiting nuclear weapons as such.[22] It is, instead, a provision requiring the pursuit of negotiations.[23] Article VI reflects a compromise between non-nuclear-weapon states that would have preferred the nuclear-weapon states to accept a legal obligation to disarm; and nuclear-weapon states that face the geopolitical reality that makes such a commitment infeasible.[24] The text, as the General Assembly eventually adopted it, requires the parties to 'pursue negotiations in good faith on effective measures relating … to nuclear disarmament.' The NPT's drafters recognized that the challenge is not to reach agreement that nuclear disarmament is the desired goal; the challenge is to negotiate practical means to achieve the goal.

The form of expression contained in Article VI is one that, in practically every other setting in which a treaty has approximated it, has been interpreted to denote an obligation of endeavor, not an obligation of result. The focus here on the endeavor accords with the understanding that has prevailed throughout the age of competing nuclear-weapon states: nuclear disarmament is an agreed *desideratum*; where effort is needed is in identifying, agreeing, and implementing effective measures to disarm. Even the ICJ, in its non-binding advisory opinion on nuclear weapons in 1996, while puzzling for its insistence that Article VI denotes an obligation of result, nevertheless took care not to overstate the scope of the obligation any further than that: even in the ICJ's reading, Article VI remains a negotiation provision, not a nuclear-weapon ban.

So, while it does not prohibit nuclear weapons, Article VI does impose a binding legal obligation. It obliges the NPT parties to pursue negotiations. Effective measures relating to arms control and disarmament are the object toward which Article VI obliges the NPT parties to pursue negotiations.

Fulfilling an obligation such as that under Article VI is not onerous, but, like any obligation, a state is not at liberty to ignore it. What must a state do, in order to fulfill the obligation under Article VI?

Self-evidently, a negotiating clause such as Article VI requires that a state come to the negotiating table.

Coming to the table, however, is not all that Article VI requires. An obligation to negotiate also requires that a state refrain from unilateral conduct that fundamentally affects the subject matter that the state is obliged to negotiate about. This part of the obligation—comprised of two essentially negative elements: a state must neither aggravate the situation that negotiations are meant to resolve nor impose its own preferred outcome—is closely related to the obligation of good faith that accompanies any obligation to negotiate.[25]

China has not come to the negotiating table, thus failing to do what Article VI requires it do: to pursue negotiations. And China continues to carry out a transformative nuclear weapons buildup, thus doing what good faith requires China *not* do: to pursue strategic *faits accomplis*[26] before the negotiating begins and to aggravate the issue about which China has committed to negotiate.[27] As I recalled in Chapter 3, an obligation to negotiate requires no concrete outcome between the states to which it applies; but it does entail these modest constraints on their behavior. China ignores the modest constraints.

China's non-compliance with Article VI

Let us consider China's non-compliance with Article VI in more detail—in particular, China's failure to come to the negotiating table; and China's nuclear weapons buildup. The buildup both threatens to impose a strategic *fait accompli* before negotiations ever begin and aggravates the problem which China has agreed it shall pursue negotiations to resolve.

China is not negotiating

Considering that NPT Article VI sets out no definition of 'negotiations,' much less an exact agenda for states to follow when they negotiate, cases might arise in which it is hard to judge whether a state has fulfilled the obligation to negotiate. China, however, has supplied signposts for measuring its conduct in regard to the affirmative element of the obligation. The signposts are visible in connection with another matter of geopolitical importance, the international relations of the South China Sea.

In the arbitration that the Philippines brought against China in 2013 regarding the South China Sea, China expressed its own understanding of the steps that a state must take if it is to satisfy an obligation to negotiate. In a Position Paper that it published on South China Sea matters, China said that 'general exchanges of view, without having the purpose of settling a given dispute, do not constitute negotiations.'[28] China notes in the Position Paper that it had carried out exchanges of view with the Philippines on 'responding to incidents at sea in the disputed areas and promoting measures to prevent conflicts, reduce frictions, maintain stability in the region, and promote measures of cooperation.'[29] According to China, such exchanges of view 'are *far from* constituting negotiations.'[30] China's contacts with the US to date regarding arms control do not go even that far. It would ring false for China to assert that China has fulfilled its obligation under NPT Article VI to negotiate, unless China were to identify interactions that go considerably further than those with the Philippines.

China's failure to meet its own standard as to the affirmative element that constitutes a negotiation is a cause for concern. However, it is in respect of the negative elements of the obligation to negotiate—the duty not to impose a *fait accompli* and not to aggravate the situation about which one is obliged to negotiate—that China's conduct should raise the most concern.

China is pursuing a fait accompli

Carrying out a rapid nuclear arms buildup in the absence of any arms control or transparency mechanism and in the absence of negotiations toward any such mechanism, China is creating a *fait accompli* that inevitably prejudices a future negotiation. It is well accepted in international practice that a state may object to an adversary attempting to impose *faits accomplis* in a setting where the state is legally obliged to pursue a negotiated outcome.[31]

This would not be the first time that a state has called attention to China's practice of attempting to impose *faits accomplis*. China's conduct in the South China Sea again supplies an illustration.

The UNCLOS Annex VII Tribunal in the *South China Sea Arbitration*, accepting the position that the Philippines had advanced in those proceedings, determined that China was attempting to impose a *fait accompli* on the Philippines and other states of the region. In particular, according to the Tribunal, 'China has effectively created a *fait accompli* at Mischief Reef by constructing a large artificial island on a low-tide elevation,' and thus intruding upon the Philippines' exclusive economic zone, an area in respect of which China evidently seeks to compel the Philippines to make concessions, notwithstanding the settled character of the Philippines' rights there.[32]

A party does not act in good faith, when, attempting materially to alter the factual circumstances to its own advantage, it delays a negotiation that it is legally obliged to pursue.[33] Nor does the party act in good faith when it demands, as a precondition to the negotiation, a negotiation on a different subject matter as to which negotiation is *not* obliged.[34] The principle of good faith is intrinsic to negotiation, including to negotiation as required under NPT Article VI. China refuses to engage on nuclear arms control in a meaningful manner, and, in the time elapsing, China continues to accelerate its qualitative and quantitative buildup of nuclear arms.[35] China's buildup alters the ground that negotiations are meant to address. As the *South China Sea* case illustrates, China's nuclear policy is of a piece with China's approach to negotiations in other strategic settings.

China's conduct aggravates the problem that China is obliged to negotiate to resolve

Under general international law, parties shall not aggravate or exacerbate a problem that they have committed to seek to resolve.[36] A number of international agreements codify the obligation; for example, Articles 279 and 300 of UNCLOS, which requires parties to refrain from 'any acts that might aggravate or extend the dispute.' However, it is an obligation of general character and, so, needs no affirmation by treaty to apply. '[S]uch a duty [to refrain from aggravation] is inherent in the central role of good faith in the international legal relations between States.'[37] It is an obligation, therefore, applicable to the conduct of all parties to the NPT.

The most rapid and sophisticated nuclear weapons buildup since the Cold War, carried out with no measure of control or transparency, aggravates the situation that Article VI obliges the parties to negotiate to resolve.

We place China's nuclear weapons buildup in context, when we consider China's recent practice concerning other problems relevant to international peace and security. Salient once again in this connection are the findings of the UNCLOS Annex VII Tribunal in the *South China Sea Arbitration* in regard to China's artificial islands. The Tribunal found as follows:

> China's intensified construction of artificial islands on seven features in the Spratly Islands during the course of these proceedings has *unequivocally aggravated the disputes* between the Parties.[38]
>
> *China has aggravated the Parties' dispute* with respect to the protection and preservation of the marine environment.[39]
>
> China has *undermined the integrity* of [the Annex VII arbitral proceedings] . . . [by having] *permanently destroyed evidence* of the natural status of [certain features in the Spratly Islands the status of which had been in dispute].[40]

The first two findings here expressly identify conduct of China that aggravated the dispute that China was legally bound to address. In the *South China Sea* case, China was legally bound to address the dispute through arbitration.[41] The principle of non-aggravation, which China violated in its dispute with the Philippines, is relevant, *mutatis mutandis*, to settings in which China is legally obliged to negotiate.[42]

The third of these findings in the *South China Sea* case—destruction of evidence—is relevant as well. Obfuscation and lack of transparency concerning the subject matter of a dispute are antithetical to negotiating the dispute. Destruction of evidence is an extreme form of obfuscation and lack of transparency.[43] China in several fields in recent years has engaged in such misconduct. China's refusal to negotiate toward transparency in regard to China's nuclear weapons buildup has troubled China's interlocutors when attempting to address Article VI in particular. China's refusal in that regard, if not 'undermin[ing] the integrity' of Article VI altogether, makes fulfilling the negotiation requirement far more difficult than it otherwise would be.[44]

China is a nuclear-weapon state, and it has an alarming record of aggravating the problems that it ought to be working to solve. It is axiomatic that, when seeking to ameliorate or resolve a dispute, states are to refrain from conduct that makes the dispute worse.[45] This is especially so when states are legally obliged to pursue negotiations. China's conduct in carrying out its nuclear arms buildup threatens irreparable harm to international peace and security. It is a clear case of a state aggravating the situation which the state is legally obliged to pursue negotiations to resolve.

Calling out the arms racer

China, through its nuclear buildup, pursues a fundamental change in nuclear deterrence—and China insists that it will negotiate only when it is satisfied that it has achieved that change. The change that China pursues is not around the margins of deterrence calculus, or even in the middle ground. If attained, the change that China pursues, instead, would upturn the existing strategic balance and equip China to assert military-political dominance in East Asia and beyond.[46] China, in this way, is in breach of Article VI—not because Article VI prohibits nuclear arms; it does not—but because Article VI requires the pursuit of negotiations in good faith; and China's buildup is incompatible with that intrinsic part of the negotiating pledge.

An Occasional Paper published by the Nonproliferation Policy Education Center in June 2022, following a diplomatic simulation, ventured that 'bureaucratic conservatism' at the US Department of State would make it difficult for the US to adopt a formal finding that China has violated NPT Article VI.[47] As an aside, it is interesting to recall that National Security Action Memorandum 335 of June 28, 1965, by which the Lyndon Johnson Administration indicated that it would pursue the nuclear arms control negotiations that eventually led to the NPT, tasked arms control to the Arms Control and Disarmament Agency—a measure which, in effect, bypassed the State Department.[48]

Whatever the bureaucratic preferences involved, a serious problem is on the horizon if we fail to reinvigorate the NPT. No provision of the NPT has more disappointed disarmament activists than Article VI, and no provision offers a more obvious path to breathing life back into the 'cornerstone' treaty of the nuclear age. A few words are in order about how to hold China to account by reference to its failure to fulfill the obligation expressed in Article VI. China's own conduct and rhetoric provide ample openings for the US and its allies to do so.

In the run-up to the 2020 Review Conference, China pronounced that it has 'never compared its nuclear-weapons investment, quantity or scope with those of other countries.'[49] This assertion is patently false—and its falsity is directly relevant to China's non-compliance with Article VI. According to China's senior arms control official, '*[G]iven the huge disparity between the Chinese nuclear arsenal and that of the US and the Russian Federation*, we simply do not believe that there is any fair and equitable basis for China to join the US and the Russian Federation in a nuclear arms control negotiation'[50] (emphasis added). So China very much has 'compared …' its nuclear forces 'with those of other countries.' Elsewhere, China states that only 'drastic and substantive reductions' by the US and Russia 'would create necessary conditions for other nuclear-weapon states to join in multilateral negotiations on nuclear disarmament'[51]—a statement that makes sense *only* as a comparison. China here expresses the matter unambiguously: China sees its obligation to negotiate under Article VI as relative and conditional. Article VI, however, is without the 'conditions' that China demands.[52]

China repeatedly has stated that it will not negotiate, as invited by the US, until China's nuclear arsenal is at parity with that of the US (and that of Russia). China's demand for parity is purposeful, and it is consequential. China makes the demand for the concrete purpose of avoiding negotiations, and by so doing China violates the binding legal commitment of Article VI to negotiate. As to China's further assertion that it 'has never participated in [a] nuclear arms race of any kind, nor will China participate in the future,'[53] this is detached from reality. There is no doubt that China has embarked upon an arms race.[54]

The comparisons and denials in China's NPT diplomacy are pervasive. 'States with the largest nuclear arsenals,' China says, 'bear special and overarching responsibilities with regard to nuclear disarmament. They should continue to drastically reduce their nuclear weapons.'[55] Practically in the same breath, China extolls 'strategic stability.'[56] How it would be that the US 'drastically' reducing its deterrent, at a time when Russia continues its invasion of Ukraine, and China threatens military force against Taiwan, would foster strategic stability, China does not say. Nor does China say how 'drastically' cutting arms on China's terms would conduce to the 'reasonable process of *gradual* and *balanced* reduction' that China says is needed.[57] Meanwhile, China's buildup of nuclear arms continues apace—the most significant threat to strategic stability and great power balance in over 30 years of nuclear-weapon-state relations.

Turning to the INF Treaty, China opposes its 'multilateralization.'[58] In connection with the INF Treaty, China repeats a familiar theme about 'maintaining global strategic balance and stability,' and again ignores that China's rapid nuclear weapons buildup is inimical to balance and to stability. China says that China's critics should stop talking about an 'exaggerated "China threat".'[59] The facts, however, are not in China's favor. China's buildup is record-setting, and it jeopardizes the strategic balance on which the world has long relied. The nuclear-weapon states agreed to Article VI in the form they did, for the reason that an immediate nuclear ban observed by some rivals but not others would endanger security. Demanding 'drastic' reductions by one or two nuclear-weapon states, while a third refuses to negotiate and races toward what it proclaims to be 'fair and equitable,' increases the very strategic risk that the original NPT states parties meant Article VI to mitigate.

China's conduct and rhetoric supply ready openings for the US and its allies to invoke Article VI. Invoking Article VI in reply to China's nuclear weapons buildup and repeated refusals to negotiate would serve two objectives at once. First, invoking Article VI would be a step toward holding China to account. Second, and just as importantly, invoking Article VI would be a step toward revitalizing the NPT.

Others have begun to notice China's failure to engage in good faith negotiations. Noteworthy in this regard, the European Parliament has voiced concern over the lack of transparency and the pace of China's nuclear weapons buildup.[60] In particular, in a resolution that it adopted in December 2021, the European Parliament noted that 'China has shown a lack of transparency and reluctance to engage in talks on its potential participation in multilateral nuclear arms control instruments, which has allowed it to stockpile unhindered a large arsenal of technologically advanced intermediate-range ballistic missiles.'[61]

In the same resolution, the European Parliament called on China 'to engage in efforts to multilateralise and universalize the successor treaty to the INF … and *to actively participate in talks about any other arms control agreements*'; and '[n]ote[d] with concern [...] the current modernization and broadening of China's nuclear arsenal, including hypersonic nuclear-capable missiles; [and] call[ed] on China to *engage actively and in good faith in international arms control, disarmament and non-proliferation negotiations*' (emphasis added).[62] Japan, too, has drawn attention to China's failure to negotiate. At the 2020 NPT Review Conference, Japan observed that China's suspected nuclear activities violate Article VI.[63]

A response thus can be seen taking shape that uses Article VI to call out the arms racer. The US and its allies should continue to use Article VI as we work to bring China to the negotiating table for nuclear arms control.

But what specific matters relating to nuclear arms control might we invite China to negotiate about?

The next chapter considers some possibilities.

FIVE

What's Left to Negotiate?

As Chapter 3 recalled, the 'pursue negotiations' clause in Article VI of the NPT indicates a legal obligation.[1] Plausibly interpreted, the obligation is that the parties in fact negotiate nuclear arms control. If parties *are* in fact to negotiate, then they must take certain antecedent steps. Some steps, though in some cases not free from difficulty, are largely ministerial—for example, choosing a venue and setting a procedural format. Others can present a fundamental challenge. In particular, the parties must identify substantive issues that they are amenable to negotiating.

In today's strategic environment, what is there left to negotiate about nuclear arms control?

The present chapter aims to invite consideration of a range of potential topics relevant to nuclear weapons that states, in particular the US and its allies, on the one hand, and China, on the other, might address in negotiations. My purpose in this one chapter of a short book is not to supply a complete analysis. It is, instead, to spur the policy debate necessary if the US and its allies are to begin the search for substantive issues that they might invite their interlocutors to place on a future negotiating agenda.

The wide ambit of Article VI

Article VI embraces a large field of potential negotiation subject matter. The breadth of Article VI is reflected in the 'effective measures' clause. The 'effective measures' clause makes clear that the parties have not confined themselves to negotiate toward one or another specific measure: the NPT does not define 'effective measures.' The NPT identifies the ultimate goal that 'effective measures' are to relate to—they are to be measures 'relating to cessation of the nuclear arms race at an early date and to nuclear disarmament'—and the NPT describes those measures to be in the *direction* of disarmament, which is to say measures forming a process of increments.[2] Prescribing no specific modalities for arms control, the NPT acknowledges the realities of geopolitical competition that motivated states to adopt the NPT in the form that they did. The NPT leaves a significant discretion to the parties when they come to identify a negotiating agenda.[3]

National policy documents, such as the NPR (US), acknowledge Article VI but have not in recent years suggested a broad range of specific topics that nuclear-weapon states might negotiate.[4] Think tanks have gone further. In an Occasional Paper published in June 2022, the Nonproliferation Policy Education Center (NPEC) suggested a number of issues that the US might invite China to address in negotiations. Among the possibilities that the NPEC paper suggested are restricting plutonium production, establishing nuclear hotlines, clarifying the terms of the Comprehensive Nuclear-Test-Ban Treaty (CTBT), and adopting new transparency measures, such as IAEA safeguarding of the most 'militarily useful' but ostensibly 'civilian' nuclear facilities.[5] A diplomatic simulation that NPEC conducted in autumn 2021 reached the conclusion that non-aligned countries are unlikely to support US efforts to place pressure on China to enter NPT Article VI negotiations, unless the US proposes serious steps toward arms control.[6] My argument in the present book largely accords with that conclusion. Not only as a matter of international law, but also international politics, the US and its allies should be prepared to negotiate, and this means considering a wide ambit of potential topics to negotiate.

My argument here is not that the US must place *every* possible topic in play for negotiation.[7] It is, instead, that prudence in maintaining strategic balance does not prevent policy makers, at least in their internal deliberations, from studying a range of possible topics. As I argued in earlier chapters of this book, the arrival of the nuclear ban treaty in 2017 introduced a new risk. That risk will grow if we do *not* negotiate. Reviving the NPT and in particular its negotiations clause should be a policy goal. We should keep the goal and the reason behind it in view as we take the measure of potential topics for negotiation and explore possible new paths to arms control.

Substantive issues for negotiation

The balance of this chapter will consider 12 substantive issues that the US and allies might consider for possible negotiation: (1) transparency; (2) confidence-building; (3) hotlines and other working-level arrangements; (4) plutonium production and fuel-cycle accountability; (5) tritium limits; (6) safeguards for fast reactors; (7) nuclear testing and definitions in the framework of the CTBT; (8) nuclear safety; (9) export controls and nuclear cooperation with non-nuclear-weapon states; (10) institutionalizing the NPT; (11) the conventional-nuclear deterrence matrix; and (12) a possible northeast Asia strategic bargain.

Overlaps and connections exist among the issues that I consider in this chapter. For example, confidence-building is naturally linked to hotlines. Fuel-cycle accountability overlaps with safeguards agreements. Transparency has relevance to most or all the others. One might group the substantive issues under a variety of rubrics. The 12 here are a convenient starting point.

A starting point, it should go without saying, is not a prescription for an end-state. The following are summary sketches. They require elaboration, and policy makers must deliberate their pros and cons before deciding whether they are fit for purpose. Moreover, even if the US and its allies were to judge that a particular matter merits negotiating, that judgment would get us only part way: for a negotiation to take place, China would have to agree to negotiate. As I have noted elsewhere in this book, a cardinal insight embodied in the NPT is that arms control and disarmament are not an isolated technical endeavor. They are entwined in geopolitics—enabled, and constrained, by the wider political and strategic setting in which nuclear-weapon states pursue them. As a former senior US arms control official recently observed, 'the question goes far beyond … issues of whether technocratic arms control experts can imagine and articulate ideas for some mutually beneficial arrangement.'[8] My aim here is neither to provide a roadmap to a new geopolitics, nor to prescribe a ready-to-use negotiating agenda.

With the caveats in mind, let us turn to the substantive issues.

Transparency

Arms control policy makers long have recognized the role of transparency in arms control.[9] The 2022 NPR (US) refers to transparency in connection with adversaries' nuclear activities.[10] It notes in particular China's 'lack of transparency' in regard to its 'nuclear expansion.'[11] However, while referring to efforts to 'enhance transparency' among the P5, the 2022 NPR does not connect those efforts directly to a renewed NPT Article VI agenda. A number of non-aligned countries and others, at the 2020 Review Conference, did make that connection, expressing the view that transparency is an important component of future efforts under the NPT.[12] The US would gain diplomatic capital if it were to pursue negotiations in regard to transparency in the Article VI frame. Moreover, if China actually lifted the veil on its nuclear weapons buildup, then the US relationship with China could move toward greater trust and stability than characterize it today.

Confidence-building

Henry Sokolski observed in 2015 that the pursuit of confidence-building measures is typically associated with arms control officials grasping for near-term activities to fill a diplomatic agenda lacking realizable long-term goals.[13] Sokolski also observed, in contrast, that 'hawkish supporters of nuclear weapons (as well as hard-headed security planners who might not be as enthusiastic about relying on nuclear arms)' focus on the long-term goals.[14] Confidence-building has a formalistic ring that excites little support outside the foreign policy bureaucracy.[15]

And, yet, we face new risks that confidence-building might help ameliorate. China's rapid nuclear weapons buildup, as well as the increase in the number of countries with nuclear fuel capabilities, has imparted instability and unpredictability to relations among nuclear-weapon states and among states that might pursue nuclear weapons. So too has the eruption of conventional armed conflict involving threats of nuclear force—that is to say, Russia's aggression against Ukraine—and the possibility of other conventional armed conflict—in particular, a future invasion or blockade of Taiwan by China.[16] Worse still, Russia's aggression has violated the fundamental understanding of the post-1945 era: countries shall refrain from use or threat of force against other countries for purposes of territorial gain.[17]

The uncertainties that result from these developments could be at least as dangerous as the Cold War. For much if not all of the Cold War, the US and USSR each understood with relative confidence what capabilities the other possessed and what geopolitical limits the other would respect. The superpowers developed informal understandings with one another, a practice that reflected and strengthened mutual restraint. (The practice drew remark from Schelling and Halperin at the time.[18]) Through confidence-building measures, geopolitical competitors today could refurbish something akin to the informal understandings that lent stability to superpower relations in the past.

The strategic environment today is one in which confidence-building measures, sensibly deployed, would not be mere time-fillers in a diplomat's diary but practical steps toward risk-reduction. Because rapid changes now are taking place in hard-to-predict directions, confidence-building is a tool that deserves a fresh look in the NPT negotiating frame.[19]

It also makes sense to go beyond traditional confidence-building and to widen the effort to address the geopolitical problems that at present impede nuclear arms control. The insight behind the Trump Administration's CEND program—Creating an Environment for Nuclear Disarmament—is that we must alleviate the security risks that weigh upon policy makers today if better prospects are to open for negotiating toward NPT Article VI goals. The Biden Administration supports the program, and a multilateral CEND meeting took place in Berlin in May 2024.[20] CEND thus has survived political differences in the US and continues to attract the engagement of allies and like-minded countries. Initiatives such as CEND and enlarging them merit consideration.

Hotlines

Confidence-building may involve relatively abstract discussions aimed at refurbishing the sorts of informal understandings that enabled Cold War superpowers to stabilize their relations. Nuclear-weapon states in the past also have reached agreement on concrete steps. Hotlines have numbered among the examples.

The US and USSR established a hotline in 1963 which has continued to operate ever since.[21] The USSR later in the 1960s established similar facilities with France and the UK.[22] India and Pakistan established hotlines with one another in the 1990s.[23] The NPEC diplomatic simulation conducted in autumn 2021 (noted earlier in this chapter) concluded that the

US should pursue, among other initiatives, improving nuclear crisis communications with China and should encourage China to do the same with its regional neighbors.[24]

A US–China presidential hotline has been in place since 1998.[25] A Defense Telephone Link (DTL) was established in 2008 to connect the two countries' defense secretaries.[26] The US Department of Defense at present calls for more US–China hotlines,[27] and 'several former senior U.S. military personnel' have been reported to say that the 'inadequacy [of US–China crisis communications systems] constitutes a clear and present danger … that could fuel a dangerous U.S.–China military confrontation at a time of heightening bilateral tensions in the Taiwan Strait and the South China Sea.'[28]

Hotlines are not a cure-all, and to use them effectively the countries that establish them must use them in good faith. Pakistan, for example, has complained of misuse of a hotline that it established with India in the 1990s.[29] Nuclear weapons policy analysts have argued that misuse is an ever-present risk,[30] because gaps exist between hotline users' intentions. A recent RAND report cautions against pursuing hotlines with China, because China intends to misuse them and, in any event, Chinese officials have failed to respond when US counterparts have attempted to reach them over the hotlines in the past.[31] The report also suggests that overtures by China's People's Liberation Army to negotiate about a hotline for crises related to outer space might not be in good faith.[32] In an earlier commentary, RAND analysts said that China intends hotlines, not to 'deconflict and de-escalate … between forces during a crisis' but, instead, to allow China to 'exploit […] crises to its advantage and manipulat[e] risk calculations.'[33] If the RAND analysts are right, then China does not intend to use hotlines to resolve crises, but 'to signal resolve, assign blame, and stall until Beijing stakes out a position of maximum pressure and leverage over the United States during negotiations.'[34]

With regard to each of the suggestions in this chapter, US policy makers—and if it comes to a negotiation with

China, US negotiators—must exercise caution. If it does come to a negotiation on hotlines, then the US should make clear that it intends any future hotline or update to existing hotlines to further the larger objectives to which the US long has committed itself under NPT Article VI. Nevertheless, hotlines merit consideration among possible topics for a new negotiating agenda.

Plutonium production and fuel-cycle accountability

China in 2017 discontinued its earlier commitment to the IAEA to report on plutonium holdings.[35] Coinciding with that retrograde step, China started work on nuclear fuel reprocessing facilities to extract plutonium from spent reactor fuel.[36] China remains the one NPT nuclear-weapon state not to formally espouse a moratorium on fissile material production. Yuki Kobayashi, a policy scholar in Japan who has worked on nuclear safeguards, suggests that '[l]ack of transparency in plutonium production and nuclear arms expansion … may emaciate the IAEA's regime for international nuclear materials management.'[37] Dr. Kobayashi is right to add that such thinning out of the IAEA's regime, in turn, would put the NPT as a whole at risk.[38] China's retreat from fuel-cycle accountability endangers the NPT. However, it also presents an opportunity for future NPT negotiations. How might the US and its allies pursue the opportunity?

One possibility is through ongoing efforts toward a Fissile Material Cut-off Treaty (FMCT).[39] China's discontinuation of reporting on its plutonium holdings, combined with the evidence of China's rapid increase of plutonium extraction, are a set-back for efforts toward the proposed FMCT. A call for fuel-cycle accountability might well meet a welcome reception from other countries. For example, the Vienna Group of Ten expresses support for the FMCT.[40]

Even China expresses its 'positive attitude' toward the FMCT.[41] So the US would not be proceeding on a blank slate if it were to raise fissile cutoff as a possible issue for negotiation. As the Biden Administration's Under Secretary of State for Arms Control and International Security has said, the time might be right 'to seek negotiations on a cut-off in the production of fissile material for use in nuclear weapons or other nuclear explosive devices.'[42]

That said, the FMCT is not the only frame in which the US and its allies might invite China to discuss fuel-cycle reporting. Given China's obstruction of FMCT discussions in the Conference on Disarmament[43] and specious linkage between fissile-material cutoff and outer-space negotiations,[44] a more promising format might be discussions among the nuclear-weapon states (P5 discussions) in the frame of the NPT itself. Attempting to negotiate with China on fissile-material cutoff, to date, has been unsuccessful.

Calling for greater accountability in the management of China's fuel cycle would be a modest, but meaningful, step. Calling for greater accountability would be reasonable, because even if China agreed merely to implement the reporting practice that it had agreed to implement prior to 2017, then progress would have been achieved against the impasse that China's retreat from accountability has caused. In other words, raising this point would not be asking China to commit to something to which China has not already committed. Instead, it would note China's commitment in 1997 to the IAEA to report on transfers of plutonium to civilian use and invite China to return to that commitment (and to observe it).

Properly communicated, well-grounded concerns about regional security would give added impetus to NPT negotiations on China's plutonium production and lack of fuel-cycle accountability. Under the Rubio–Markey Act of 2016, it was proposed that Congress find, *inter alia*, that 'China's plans to reprocess United States-origin spent fuel raise the risk that China could divert plutonium to military purposes, enabling it to produce additional nuclear weapons that threaten the United States and its allies' and that 'China's pursuit of commercial plutonium reprocessing will increase the likelihood that Japan expands its commercial reprocessing program and that South Korea will increase efforts to initiate a similar program of its own.'[45] The risk of such regional escalation is one that China itself is best positioned to address. Fuel cycle accountability and limits on plutonium production would reduce the risk. As I will suggest at the end of this chapter, the US also might propose connecting accountability and plutonium limits to a possible wider bargain in respect of nuclear weapons in northeast Asia.

The 2022 NPR calls for China to 'adopt a moratorium on fissile material production or, at a minimum, provide increased transparency to assure the international community that fissile material produced for civilian purposes is fully accounted for and not diverted to military uses.'[46] However, as is the case with its calls for transparency in general in regard to China's nuclear weapons buildup, the 2022 NPR does not connect fuel-cycle accountability to a renewed NPT negotiating agenda, or at least not in clear enough terms to contribute to the revitalization of the NPT that I argue would serve the strategic interests of the US and its allies to pursue. Calling out China for its radical increase in fissile-material production capacity, as the NPR does, is necessary, but the US also should be calling China to the table to negotiate.

Tritium limits

Tritium—the rare and rapidly decaying isotope of hydrogen containing one proton and two neutrons[47]—is a byproduct of nuclear power generation.[48] Tritium routinely attracts the attention of regulators addressing its effects on human health and the environment.[49] Tritium has gained little attention in respect of non-proliferation. However, in a detailed treatment of the matter in 2002, Kenneth D. Bergeron, who had worked for 25 years at Sandia National Laboratories on commercial nuclear reactor safety and military production of tritium, warned of 'a relaxation of concern about the connection between tritium and weapons proliferation.'[50]

There is no doubt that tritium is relevant to nuclear weapons. The isotope is critical in 'boosting' the yield of the fission stage in hydrogen bombs.[51] Fission bombs themselves rely on tritium to increase their efficiency.[52] In the supply chain for building a nuclear weapon, tritium is an important link.

Moreover, because it decays rapidly, tritium is a link with a distinct vulnerability. Estimates are that 5.5 percent of the tritium in an atomic weapon is lost through decay every year.[53] Unlike the main fissile materials, plutonium and uranium-235, which have extremely long half-lives (that is to say, they decay very slowly), the tritium in an atomic weapon needs to be replenished, and so a nuclear weapons program needs continued access to tritium. On these grounds, the supply of tritium to a nuclear weapons program merits consideration as a possible choke-point which a non-proliferation strategy might target.

Like any control on materials or technologies needed to build a nuclear weapon, control on tritium will have stronger non-proliferation effects, the more tritium-producing states accept and enforce the control. The US already has export controls on tritium and technology necessary for its production.[54] The US should consider inviting China to negotiate a possible fresh initiative to strengthen international controls on this proliferation-sensitive isotope and to improve China's dubious nuclear export record in general.

Safeguards for fast reactors and keeping the IAEA in the picture

IAEA safeguards under NPT Article III 'are fundamental to the nuclear non-proliferation regime and help create an environment conducive to nuclear cooperation.'[55] Even though IAEA safeguards are no panacea,[56] and safeguards have little effect unless *national* authorities implement them ably and in good faith, the continued vitality of the NPT requires the continued observance of NPT Article III.

As I recalled in Chapter 1, Article III formally addresses only the non-nuclear-weapon states. The nuclear-weapon states long have stated publicly that they appreciate, however, that Article III's formal addressees—the non-nuclear-weapon states—gain confidence in the IAEA safeguard system when they, too—the nuclear-weapon states—account for their nuclear activities.

The need for reliable safeguards on non-nuclear-weapon states' nuclear activities has grown, as the number of states has grown that either make enriched uranium, reprocess spent reactor fuel, or claim that they are at liberty to do so. A confident estimate some years ago already counted among these Japan, South Korea, Argentina, Brazil, South Africa, Egypt, Turkey, Saudi Arabia, Iran, Vietnam, and Jordan.[57]

To get a sense of the difficulties that would arise if more non-nuclear-weapon states were to pursue reprocessing, it helps to observe that fully 20 percent of the IAEA budget in 2010 was allocated to monitoring the one plutonium reprocessing plant in a non-nuclear-weapon state—Japan's reprocessing plant at Rokkasho.[58] The impact that reprocessing has on the Article III safeguards system has long been evident. In this light, too, China's rapid buildup of plutonium processing capacity is cause for concern.

In 2019, China boasted in one of its NPT Preparatory Committee reports that it has 'actively reduced the use of sensitive nuclear materials and has completed the decommissioning of two domestic research microreactors and a low-enrichment retrofit project for a research microreactor.'[59] China also mentioned its 'demonstration fast-reactor project.'[60] China did not say anything about IAEA access to that project. In the same Preparatory Committee report on NPT implementation, China claimed that 'China adheres to the "closed cycle" model of nuclear fuel development and has essentially completed its nuclear fuel cycle system, with the supply of nuclear fuel able to meet the fuel requirements of the nuclear power plants that have been put into operation.'[61] As an account of China's nuclear fuel activities, this claim is, at best, selective. China does not acknowledge the real issue: China is massively expanding its capacity to produce fissile material.

Negotiating for a more rigorous safeguarding of China's nuclear activities would be no mere cosmetic exercise. It is likely that China already reprocesses fuel from nuclear power plants and uses civil reactors to produce tritium. Such activities could violate China's peaceful end-use agreements with the US and arguably, too, would violate China's 1997 understanding on voluntary reporting with the IAEA.[62] Short cycling and reprocessing of spent fuel from China's US AP 1000 reactors and Westinghouse-designed Chinese variants has raised justified concern in the US Senate.[63]

The US might explore, in return for meaningful steps by China, a pause on commercial deployment of fast reactors and spent fuel recycling in the Pacific Rim, a natural step taking its lead from the Rubio–Markey Act.[64] The US might also propose—or one or more non-nuclear-weapon states might propose—that all nuclear-weapon states open their civil nuclear facilities to IAEA inspection.[65] The US long has been the leader among the NPT nuclear-weapon states when it comes to transparency. The US should consider ways to leverage its strength in that area to revitalize the negotiating agenda under NPT Article VI.

The US also should continue to implement measures to ensure that proposed advanced modular fast reactors and recycling plants operate under safeguard agreements. Non-proliferation officials during the Trump Administration pursued a number of helpful steps toward safeguards for new classes of civil nuclear power infrastructure. For example, under the FIRST program (Foundational Infrastructure for Responsible Use of Small Modular Reactor Technology), which the Biden Administration has continued, non-proliferation officials engage with foreign partners to lay the groundwork for non-proliferation-compliant civil nuclear energy projects.[66] Under the Trump Administration and Biden Administration, the US also has concluded a number of Nuclear Cooperation Memoranda of Understanding (NCMOUs)—non-binding bilateral agreements that set the stage for civil nuclear energy with foreign partners in a framework that will implement safety, security, and non-proliferation standards.[67] Programs such as these demonstrate the US's good faith commitment to the Article IV guarantee of access to peaceful nuclear energy. They also demonstrate that the US takes its non-proliferation obligations seriously. A renewed negotiating agenda under Article VI could aim to ensure that China takes its non-proliferation obligations seriously as well. Pursued in good faith by the US and China, a frank discussion of safeguards would increase the nuclear-weapon states' credibility when they remind non-nuclear-weapon states that Article III remains a vital part of the non-proliferation regime.

Nuclear testing—let's talk definitions (CTBT framework)

The US Congressional Research Service notes that '[a] ban on nuclear testing [is] the oldest item on the [nuclear] arms control agenda.'[68] India made an early proposal (in 1954) for a test ban treaty; tripartite negotiations (UK, US, and USSR) started in Geneva in 1958 in a Conference on the Discontinuance of Nuclear Tests.[69] After those negotiations ran aground, the Conference closed. Negotiations resumed in March 1962 in the UN Disarmament Committee.[70] As the Introduction to this book recalls, the three parties concluded a Partial Test Ban Treaty in 1963.[71]

Negotiations toward a comprehensive ban on nuclear testing have been associated with the NPT since the NPT's inception. As the time approached in the early 1990s to decide whether to extend the NPT indefinitely,[72] NPT non-nuclear-weapon states were disappointed that the nuclear-weapon states had not yet adopted a comprehensive test ban. They argued that the failure to adopt a comprehensive test ban underscored a fundamental unfairness in the Treaty. Non-nuclear-weapon states said that they had given up a great deal by supporting the NPT—what they said otherwise would have been a sovereign right to acquire nuclear weapons; but (they said) the nuclear-weapon states had given up nothing. The non-nuclear-weapon states argued, in short, that the time had come to remedy an imbalance in the NPT bargain. As I suggested in Chapter 1, the bargain that those states accepted when they became party to the NPT is the bargain reflected in the text of the NPT; nothing in the NPT legally obliges the nuclear-weapon states to adopt a test ban. However, a future ban on nuclear tests, in the eyes of the NPT non-nuclear-weapon states, came to be 'the touchstone of good faith on these matters.'[73]

At the NPT Review and Extension Conference of April–May 1995, non-nuclear-weapon states insisted that the nuclear-weapon states make a material commitment to offset the perceived unfairness in the NPT bargain. As a result, the Conference elevated the achievement of a test ban to a matter of priority. The Final Document of the 1995 Review and Extension Conference connected the achievement of a test ban treaty expressly to NPT Article VI. According to the Conference Final Document: 'The achievement of the following measures is important in the full realization and effective implementation of article VI … (a) The completion by the Conference on Disarmament of the negotiations on a universal and internationally and effectively verifiable Comprehensive Nuclear-Test-Ban Treaty no later than 1996.'[74]

The Final Document further described the prospective ban as helping nuclear disarmament be 'fulfilled with determination.'[75] One might object that, because Article VI does not oblige any specific arms control agreement, the Conference should have described the link to Article VI in more restrained terms than 'full realization and effective implementation.' However, if we read the 1995 Final Document to have proposed a comprehensive test ban treaty as a *desideratum*, not an obligation, then the proposal was well within the compass of Article VI.

Negotiations toward a comprehensive test ban treaty[76] took place in the frame of the UN Conference on Disarmament (CD) and resulted in a draft,[77] which Australia duly submitted to the UN General Assembly.[78] The UN General Assembly adopted the CTBT on September 17, 1996 and called on 'all States to sign and ... become parties to' it.[79] Australia's engagement on the CTBT in the General Assembly and the support that other US allies gave to the drafting process in the CD[80] illustrate the importance these states attach to the CTBT. Indeed, some 20 years later, as the 2020 NPT Review Conference approached, US partners, allies, and like-minded countries continued to lay emphasis on the CTBT, describing it, for example, as an 'integral part of the 1995 decision to indefinitely extend' the NPT and 'a core element of the nuclear disarmament and nuclear non-proliferation regime.'[81] At the time of the CTBT's adoption, non-aligned countries also expressed their support,[82] which they have continued.[83]

China, at the time of the CTBT's adoption, while supporting the CTBT, said that the CTBT 'is not entirely satisfactory.'[84] China sought to justify its misgivings by referring to what the CTBT does *not* do: 'it does not touch upon the conclusion of a convention on the comprehensive prohibition of nuclear weapons.'[85] Yet China has joined the other nuclear-weapon states in rejecting the TPNW, a treaty that certainly *does* 'touch upon' a comprehensive prohibition. Insisting on a comprehensive prohibition against nuclear weapons is part of China's diplomatic repertoire, but China is careful not to subject itself to such a prohibition or even to a more modest obligation such as the test ban. It is telling in its internal contradiction that one of the NPT nuclear-weapon states *least* likely to be constrained by the suasive effects of the nuclear ban treaty also says that the absence of a nuclear-weapon ban prevents it from concluding realistic nuclear arms control agreements. As I noted in Chapter 2, the prohibitionist agenda has asymmetric effects between states that have open political systems and those that do not.

The US signed the CTBT on September 24, 1996.[86] The US Senate rejected the treaty in 1999.[87] China and the US are the two NPT nuclear-weapon states not to have ratified the CTBT; Russia, though having ratified, declared on November 2, 2023 that it was withdrawing from the treaty.[88] Other states, too, among the 44 whose ratifications are needed before the CTBT enters into force have not ratified.[89] Russia considers itself no longer bound by the CTBT and probably carries out nuclear weapons tests,[90] and China conducts experiments at the Lop Nur nuclear weapons test site that also likely exceed CTBT constraints.[91] The US Senate's continued hesitancy to consent to ratification of the CTBT is well grounded.[92]

Partners and allies of the US, as well as non-aligned countries, lament that the CTBT still has not entered into force. For example, in the Preparatory Committee for the 2020 NPT Review Conference, countries comprising the 'Non-Proliferation and Disarmament Initiative' (Australia, Canada, Chile, Germany, Japan, Mexico, the Netherlands, Nigeria, the Philippines, Poland, Turkey and the United Arab Emirates) expressed their 'strong concern' that over 20 years had elapsed since its adoption without the CTBT having entered into force.[93]

Efforts to advance the substance of the CTBT would demonstrate that the US recognizes the concern that these countries and others have expressed.

However, a crucial element of unsettled substance—and persistent stumbling block to the CTBT—is the definition of 'nuclear test.'

The 2020 US Arms Control Compliance Report[94] observes that '[a] key challenge in interpreting adherence to [voluntary testing] moratoria is determining what each moratorium actually means.'[95] The US takes the view that a commitment not to conduct 'nuclear explosive tests' entails a 'zero-yield' standard.

Contested definitions long have led to obstruction and delay in arms control.[96] Strategic offensive systems that Russia has newly developed offer a conspicuous example. The 2022 NPR observes that 'Russia is pursuing several novel nuclear-capable systems designed to hold the U.S. homeland or Allies and partners at risk.'[97] These systems include Russia's Burevestnik nuclear-powered cruise missile, Kinzhal air-launched ballistic missile, and Status-6 Poseidon nuclear torpedo. Then-US Under Secretary of State for Arms Control and International Security Andrea Thompson stated to Congress in 2019 that, for purposes of the New START Treaty, the US considers the expression 'strategic offensive arms' to embrace these three systems.[98] Russia insists that the systems 'have nothing to do with the strategic offensive arms categories covered by the [New START] Treaty.'[99] According to Russia, it is 'inappropriate to characterize new weapons being develop by Russia that do not use ballistic trajectories of flight moving to a target as "potential new kinds of Russian strategic offensive arms".'[100] Motivating the skirmish over New START definitions, the boundaries of a definition can determine whether or not particular conduct, or a particular weapon system, falls within a treaty prohibition or limit. So definitions matter.

China prides itself on its role in the P5 Working Group on the Glossary of Key Nuclear Terms.[101] China also continues to profess support for 'the purposes and objectives' of the CTBT.[102] Yet significant ground remains to be covered, before working definitions relevant to nuclear weapons tests are agreed.[103] Without agreement as to definitions, all the technical challenges of verification and compliance in regard to a test moratorium are compounded. Even in a situation where all negotiating parties proceed in good faith, they must resolve semantic differences if they are to conclude a meaningful agreement.

And the obstacles to participation in the CTBT are not limited to semantics. The 2020 US Arms Control Compliance Report refers to evidence that China might be conducting nuclear explosive tests. The Report says that Russia 'has conducted nuclear weapons experiments that have created nuclear yield.'[104] Lack of transparency as to those states' nuclear activities presents a further obstacle to agreeing to a comprehensive test ban. The matter nevertheless is one that the US might consider pursuing as a further arms control negotiating topic.

Nuclear safety

With China's civil nuclear generating output fast approaching that of France, the country long holding the record,[105] specialists have raised questions about the safety and governance of nuclear power in China. Andrews-Speed, for example, a Senior Principal Fellow at the Energy Studies Institute at the National University of Singapore and a Senior Research Fellow at the Oxford Institute for Energy Studies, says that areas of concern in China's civil nuclear industry include lack of capacity and of independence in China's National Nuclear Safety Administration; an incoherent legal framework for nuclear safety and security; deficient provision for civil nuclear liability, including in regard to international effects of nuclear accidents; and inadequate public consultations when it comes to the planning of new power plants.[106]

NPT parties have addressed nuclear safety in the civil sector. The Vienna Group of Ten, for example, in connection with the 2020 Review Conference, included under the nuclear security rubric environmental safety and notification of coastal states when a state plans to transit nuclear materials.[107] Similar issues might merit inclusion as well in a revitalized negotiations agenda under Article VI. These issues are relevant to the serious concerns arising from evidence that China is diverting nuclear fuel from its civil reactor program to weapons use.[108] To be sure, nuclear safety relies chiefly on national regulatory régimes and enforcement. Nevertheless, the impact of safety failures is potentially international. The US and other negotiating partners thus can credibly refer to their concerns over China's safety rules and practices in the frame of Article VI.[109]

Export controls and nuclear cooperation with non-nuclear-weapon states

Experts have warned that China, in the pursuit of export markets in emerging technologies, ignores distinctions between civil and military end-use, thus frustrating one of the main policy objectives of export controls.[110] As for illicit acquisition of nuclear materials and equipment for its own use, China's misconduct is well established.[111]

A number of US allies and like-minded countries have identified nuclear export controls as an area meriting attention in the frame of the NPT. For example, in a joint paper submitted during the Preparatory Committee meetings leading up to the 2020 NPT Review Conference, the Vienna Group of Ten stated, 'Nuclear export controls are a legitimate, necessary and desirable means of implementing the obligations of States parties under Article III.'[112] *Mutatis mutandis*, export controls would seem a possible topic that at least some NPT parties would welcome adding to a future agenda for negotiations in the frame of NPT Article VI. The troubling admixture by China of civil programs and military applications, including through its Military-Civil Fusion doctrine,[113] further suggests that good faith negotiations on export controls in the nuclear field would be a fruitful step for arms control and non-proliferation.

Nuclear cooperation with non-nuclear-weapon countries has a natural relationship to export controls. How the US and China, as nuclear-weapon states, might manage their cooperation with non-nuclear-weapon states on nuclear matters is a related area that the US might consider inviting China to explore as a possible negotiating topic.

To remain a credible interlocutor with China on nuclear cooperation with third countries, the US would have to take care that its own nuclear cooperation with third countries continues to meet the highest standards. Calling on China to adhere to a more responsible approach to nuclear cooperation than it pursues today, the US would weaken its credibility if it were to relax its own nuclear cooperation policy.

For an example of US nuclear cooperation policy at work, the US can refer to its relationship with the United Arab Emirates (UAE). US–UAE nuclear cooperation takes place under the so-called 'gold standard' of non-proliferation. When the UAE entered a nuclear cooperation agreement with the US in 2009,[114] the UAE agreed to forgo enrichment and reprocessing, and it committed to enhanced inspection and transparency. An Additional Protocol between the UAE and the IAEA now embodies the UAE's agreement in this regard.[115]

Though the 'gold standard' is a long-standing feature of US nuclear cooperation policy, it is necessary at times to recall its importance and the difficulties that a relaxation of the standard would entail. In a joint letter in September 2023, a number of former non-proliferation officials, including myself, urged the President of the United States to maintain the highest non-proliferation standards in US dealings in particular with Saudi Arabia.[116] Saudi Arabia is a non-nuclear-weapon NPT country that seeks to expand its civil nuclear power industry but that also has alluded to the possibility of pursuing nuclear weapons.[117] If taking a more relaxed approach to nuclear cooperation, then we would have to ask what proliferation risks that approach entails, and we would have to ask what effect it has on non-proliferation at large.

It would be worth exploring with China a joint commitment to the 'gold standard' or something equivalent to it, together with meaningful verification and compliance measures to ensure implementation of the commitment in practice. Placing the matter on the NPT negotiating agenda would be noteworthy, as agenda items are typically confined to the relations of the negotiating parties. However, nothing in the NPT bars exploring as a possible negotiating topic the negotiating parties' nuclear cooperation with third countries.

Negotiating about third-country cooperation might open the door to a new approach to ensuring observance of NPT Article III. The goal would be to set up guardrails around nuclear cooperation, so as to keep nuclear export competition within the limits which all NPT parties are obliged to respect. The US, with its 'gold standard' approach, long has led the world in respecting the limits. Inviting China to commit to a similar approach would test China's willingness to pursue meaningful steps in accordance with the Article VI duty to negotiate.

Institutionalizing the NPT?

The Chemical Weapons Convention, the Biological Weapons Convention, and the Comprehensive Nuclear-Test-Ban Treaty each has its own permanent organization staffed and funded to provide a focal point for diplomatic engagement. The respective treaty organizations also provide a framework in which the parties may seek solutions in the face of credible evidence of non-compliance.[118] The NPT, though kindred to these other arms control treaties, has no such institutional expression. NPT Article V envisaged 'an appropriate international body' but for the limited purpose of sharing the supposed benefits of 'peaceful ... nuclear explosions.' (There is no peaceful explosions body.) The NPT is silent as to whether the parties might establish a general-purpose NPT body. It neither requires them to do so nor prevents them.

Henry Sokolski, in a retrospective on the NPT in the American Bar Association's National Security Law forum in 2023, suggested that a permanent NPT administrative body would assist in implementing the treaty.[119]

The standing organizations constituted under the other major arms control treaties suggest how an NPT organization might work, but there are pitfalls. An NPT organization all too readily could invite further competing interpretations of the treaty, without increasing the chances of resolving differences or advancing shared projects. There is also the additional financial outlay needed to establish and run a new international organization.

Moreover, institutionalizing a treaty is no assurance that the treaty will work. Article X of the Open Skies Treaty provides for an Open Skies Consultative Commission, a body intended 'to promote the objectives and facilitate the implementation' of the Treaty. The Open Skies Treaty nevertheless ceased to function reliably. The US withdrew from the Open Skies Treaty on November 22, 2020; Russia on January 15, 2021.[120]

A proposal for a general non-proliferation body would not be an entirely new idea. Suggestions to constitute such a body can be found throughout the history of nuclear arms control.[121] Yet every treaty is different, and how parties interpret and apply a treaty can change over time. In considering potential topics for negotiation, the US and its allies, if approaching the matter with realistic expectations, might explore institutional formats for administering the NPT similar to those employed in other areas but adapted to the needs of the 'cornerstone' of nuclear arms control.

The conventional-nuclear deterrence matrix: strengthening the whole by moderating reliance on the nuclear part

China says that nuclear-weapon states should 'diminish the role played by nuclear weapons in their national security policies.'[122] In drawing attention to the conventional-nuclear deterrence matrix, China opens the door to a potentially meaningful line of negotiations: China's increasing threats of conventional armed force in the Pacific, not least of all against Taiwan and the Philippines, merit discussion (and, of course, action[123]). The US should consider raising the matter in the NPT Article VI frame.

Pertinent here, success in negotiating other nuclear issues could enhance security overall. For instance, negotiating fissile-material accountability with China would demonstrate commitment to NPT Article III which, in turn, would add credibility when the US, allies, and like-minded countries call on others, such as Saudi Arabia, to give responsible accounts of their own civil nuclear projects.[124] Enhancing conventional deterrence for the benefit of countries that otherwise entertain the pursuit of nuclear weapons would have synergy, in turn, with renewed efforts under Article III.[125]

It is expensive to build nuclear weapons. A shift of the deterrence matrix toward nuclear weapons is likely to divert fiscal appropriations from other efforts, including advanced conventional weapons (which might supply more security for less money, or so it might be argued[126]).

Battlefield evidence from Ukraine and elsewhere seems to confirm that US investments in advanced conventional weapons have been well placed (even if by no means adequate for the contingencies for which the US and its allies together must be prepared). Negotiations toward possible nuclear weapons limitations, if the negotiations succeeded, would allow the US to moderate its reliance on the nuclear dimension of deterrence. China itself opens the door to negotiations on this matter, when it says that it would like nuclear weapons to play a diminished role in security. The US should consider how it might use NPT Article VI negotiations to put China to the test in regard to the nuclear part of the security equation.

A northeast Asia strategic bargain?

In addition to the other possible negotiating topics that I have outlined, the US might consider a larger strategic proposition: mutual concessions of major equities between the US and China in northeast Asia. The concessions could take a variety of forms. For example, the US could propose a moratorium on the deployment of US nuclear weapons in northeast Asia—in return for China implementing a rigorous and verifiable moratorium on China's plutonium production. To give another example, the US could propose, in return for the same or similar moratorium by China, to seek commitments from regional US allies, in particular Japan, not to pursue nuclear weapons. Any such proposals would include verification mechanisms, and no such proposals would be advanced before careful study of their costs and benefits. (The same caveats apply, *mutatis mutandis*, to every potential NPT negotiating topic.)

Mixed bargains—under which a party trades freedom of action in one domain for a verifiable commitment from the other party elsewhere—run against the grain of US arms control practice. It was the USSR during the Cold War that sometimes proposed apples-and-oranges arms control agreements, and it is China today that proposes exchanging concessions in disparate fields. For example, China repeatedly advances the suggestion that the US enter into a space treaty as part of a bargain over nuclear weapons.[127] There are reasons for caution about a mixed bargain, not least the difficulties in forecasting the effects of concessions in unlike categories. Achieving such bargains is all the more difficult, where the bargains involve the interests of third parties—in particular, allies of one of the principal negotiating parties.

Yet the act of negotiating requires each party to put forward novel ideas, in particular, ones that depart from the party's existing preferences and proposals. Repeating proposals after the other party has made clear its lack of interest in them is a dead end.

In northeast Asia, China's rapid nuclear weapons buildup alarms defense policy makers in Japan and South Korea. Talk in Japan, in particular, of a nuclear deterrent independent of the US no doubt raises concern in China, all the more so in light of Japan's extant plutonium stockpile. In northeast Asia, fault lines in the strategic status quo are readily observable.

China might recall, too, that in the US, policy makers have had earnest debates for decades about the strategic benefits of allies and others developing nuclear weapons. In the 1960s, before the Johnson Administration committed to pursue a non-proliferation treaty, some policy makers argued that acquisition of nuclear weapons by Japan and West Germany, and by India as well, would advance US security interests.[128] Shortly after he took office as president, his advisors told Richard Nixon that 'independent nuclear weapons capability [in the hands of some allied countries] might be desirable.'[129] History suggests that the possibility of inter-allied proliferation can motivate a great power to take non-proliferation more seriously than it otherwise might. Francis Gavin, a historian of nuclear proliferation and the Cold War, suggests that the prospect of West Germany acquiring independent control over nuclear weapons spurred the USSR to make its first proposals for non-proliferation.[130] If China is concerned about regional proliferation, then it is reasonable for the US to ask what concessions China is prepared to make in return for the US putting its shoulder to the wheel to dissuade US allies from seeking nuclear weapons.

★ ★ ★

As with each of the topics that I have sketched in this chapter, a great deal of elaboration would be needed before the US could consider exploring a northeast Asia strategic bargain. To emphasize again, the inventory here of 12 possible topics for negotiation is meant as a starting point for US and allied policy-making discussion, not as a prescription for immediate diplomatic overtures. Framing the discussion, however, we should keep in view the current drift away from realistic arms control. We should keep in view the risk if we fail to revive arms control under the NPT.

Conclusion: An NPT Future and Bringing Realists Back to Arms Control

The NPT today is an object of ambivalence among realists. It has not always been, however. At the start, realists had the insight that stability is the *sine qua non* of security in an age of nuclear-armed powers, and the realists' insight inspired states to draft the NPT the way they did. States drafted the NPT to avoid a worst-case outcome—an escalating rivalry between nuclear-armed states from which they would never escape. Article VI of the NPT, by requiring that states pursue a negotiated settlement of the challenges of nuclear arms, ensured that states preserve the chance to arrest the rivalry. By placing the emphasis on the elusive goal—*effective* measures to achieve arms control—Article VI, moreover, pledged states to negotiate arms control agreements that might actually work in the world as the world exists.

Realists resiled from the NPT when they began to perceive that the prescriptions that disarmament activists were drawing from the NPT would impede a rational and security-minded approach to nuclear arms control. The NPT, however, is no longer the only treaty that affects nuclear arms control.

We live now in a world of *two* treaties. On January 21, 2021, the eve of entry into force of the TPNW, the Lord Bishop of Coventry speaking in the UK House of Lords said that 'as of tomorrow the TPNW will be no less a reality for the UK than for countries that support it.'[1] He forecast that the TPNW's 'underlying humanitarian motivations will loom large over any future discussion of our non-proliferation responsibilities.'[2] As a matter of international law, to a country that is not party to a treaty, the treaty *is* less a reality than to a country that is, but the Lord Bishop was not making a lawyer's point. The TPNW has attracted the support of the disarmament community and of many governments. If the support continues to grow, then the TPNW soon will be the focal point of discourse on nuclear weapons. It will 'loom large' over any effort to revive negotiations on nuclear arms control.

I have argued in this book that we should take the TPNW seriously, for it augurs a new era in security diplomacy. A TPNW era will not be congenial to the interests of the US, its allies, or like-minded countries. Security diplomacy must address the realities of nuclear deterrence in face of great power competition and a rapid nuclear weapons buildup by the chief competitor, China. The TPNW is not about the realities. It is about the more ardent hopes of disarmament activists, and those hopes matter, because they reflect shared aspirations for security in a dangerous world; and, moreover, because activists exert influence on governments answerable to electorates in free societies.

The TPNW, though unrealistic in its aims, very much affects reality. The influence of the TPNW is not symmetric between free societies and those of our geopolitical competitors, and, so, the TPNW—if it were to become the focus of international efforts toward disarmament—would impede our efforts to maintain an effective deterrent. It does not take flights of imagination to see how this might happen, for something similar has happened before.

I said in the Introduction that this is a book about the *future* of nuclear arms control, not its past, but here in the Conclusion, too, we should remind ourselves of a salient experience from the past. In the 1980s, in what has been called 'one of the last and most controversial acts of military containment by NATO during the Cold War,'[3] the US deployed Pershing II and Ground Launched Cruise Missiles in Europe. Nuclear ban activists pushed the US to withdraw the missiles and US allies to expel them.[4] If the US or its allies had done as the activists wished, then the resultant imbalance between the Western Alliance and the Warsaw Pact would have placed international security at risk. No credible Western capability would have remained in answer to the USSR's intermediate-range ballistic missiles, the SS-20s. Ban activism at that time, if it had attained the activists' aim, would have destabilized Europe and the world.

The activist effort in the 1980s was no marginal event in European politics. In West Germany in particular anti-missile activism was a central challenge in national politics.[5] Fortunately, political leaders, not least among them Margaret Thatcher and Ronald Reagan, maneuvered the cruise missile issue toward negotiations with the USSR,[6] so that the eventual drawdown of this category of weapons took place in a more or less symmetric fashion under the INF Treaty concluded in 1987 between the US and USSR.[7] It took political leadership and adroit negotiation to avoid a destabilizing unilateral retreat.

The drama of the intermediate-range nuclear missile ban unfolded scarcely 30 years after World War Two. Memories were fresh in Europe of brutal dictatorships, aggression, and genocide. Perhaps Russia's invasion of Ukraine is reminding European allies and partners that the world remains a dangerous place, but the evidence so far is that not everyone is taking the lesson to heart.[8]

Russia's war of aggression against Ukraine should worry a wide range of governments, not just those in Europe. It especially should worry elites in countries whose neighbors harbor territorial ambitions of their own. It was not without reason that leaders in Africa and South Asia in the 20th century placed considerable emphasis on the permanence of agreed boundaries, and leaders in Latin America, the century before, the same. All the more striking that non-aligned and less-developed countries recently have shown reticence over global security, even in the face of Russia's warmaking. These countries constitute an important geopolitical interest. Among some of them, the TPNW enjoys considerable support. As they did during the Cold War, they present themselves as a counterpoise to a security order led by the US, but the economic growth of Brazil, South Africa, Indonesia, and others imparts more credibility to such a grouping than it had before. We should not take a turn toward the TPNW by any of these countries lightly.

Even where the TPNW so far has not been embraced, a shift is detectable in its favor. I have drawn attention in this book to the examples of Australia at the UN General Assembly and of Labour politicians in the backbenches at Westminster. Across Europe and elsewhere, others are beginning to move the same way. A new generation of disarmament activism is gaining influence. The activism is not limited to idiosyncratic corners of Western politics or to non-aligned countries long accustomed to offering themselves as an alternative to the main poles in international affairs. Support for the TPNW is broadly based, and the treaty will have profound effects on security diplomacy if its influence on governments continues to grow.

As the disarmament community shifts its efforts to the TPNW, foreign policy leaders in the US so far have taken only halting steps to restore confidence in the NPT. In this book, I have considered the 2022 Nuclear Posture Review and its treatment of the NPT. It is good that the 2022 NPR has spoken clearly about what the TPNW does not do: the TPNW does not provide a framework in which nuclear-armed rivals will realistically pursue safe and effective measures toward arms control and disarmament. However, national security strategists must be clear about what the TPNW *does* do: it sends the disarmament discourse down a blind ally of absolutism, similar to that which security strategy faced in the 1980s from the movement for a cruise missile ban. We need to do more, if we are to address unrealistic approaches to arms control as embodied in the TPNW and ameliorate their effects on the strategic stage. It is in this light that we must revitalize the NPT.

We must revitalize the NPT to avert the danger to which its fading from the scene gives rise. If the NPT fades away altogether, then a vacuum will result that the TPNW will fill. Avoiding that state of affairs will not be easy. As I have suggested in the preceding chapters, disarmament activists and nuclear security 'hawks' partake, by turns, in indifference and antipathy toward the NPT. The NPT has disappointed the activists, because it has not led to the complete disappearance of nuclear weapons. The NPT either holds no interest for defense-minded foreign policy thinkers, because its Article VI negotiating clause seems too abstract to mean very much, or it arouses their mistrust, because, though it originated in realists' insight about how nuclear deterrence works, it long served as rally point for the disarmament community. This is why the NPT today is almost an orphan on the US foreign policy and defense landscape.

And, yet, though this book has drawn attention to the near abandonment of the NPT, we still find traces of support for the NPT in divergent corners of US foreign policy thought. As I wrote with Henry Sokolski in August 2022,[9] the Biden Administration evidently agrees with the Trump Administration that China's refusal to join nuclear arms control talks is cause for concern, arguably a breach of NPT Article VI.[10] As noteworthy as is the Biden Administration's public affirmation that China threatens the stability that long had prevailed among nuclear-weapon states, noteworthy too is the readiness of staunch exponents of nuclear deterrence to invoke the NPT. Marshall Billingslea, who served as Special Presidential Envoy for Arms Control in the Trump Administration and more recently as one of the Commissioners of the Congressional Commission on the Strategic Posture of the United States, declared that 'China is perilously close to standing in direct violation of the NPT.'[11] Together with his then-principal, US Secretary of State Mike Pompeo, Ambassador Billingslea drew further attention to Article VI in particular when calling out China for its nuclear weapons buildup.[12] US Senator Ted Cruz, among the staunchest exponents of nuclear deterrence, is largely in concurrence with ICAN, the International Campaign to Abolish Nuclear Weapons, when it comes to the importance of negotiations under Article VI.[13] That people and institutions who approach arms control from such different starting points arrive at the same conclusion suggests the merit in placing renewed emphasis on the NPT. It also suggests that there is hope for bringing about the revival for which I have argued in this book. Many are estranged from the NPT, but the treaty is not yet *entirely* an orphan.

We need a determined effort once more to embrace the NPT. The voices audible across the political spectrum in the US supporting the NPT evince willingness to lead the effort. However, so far, we hear little if any discussion of practical steps. In this book, I have argued that NPT Article VI, the negotiations clause, provides impetus for us to enter the discussion in earnest.

Article VI, a touchstone through nearly 60 years of arms control diplomacy, exercises a sentimental attraction for diplomats. But the elusiveness of its terms is frustrating for policy makers. When policy makers seek to define a course of action that the US and its allies can confidently pursue, they have difficulty knowing what to make of Article VI. This is a negotiation provision identifying a distant end-point—disarmament. It tells us little or nothing about how to get there. The very word 'disarmament,' so frequently used in absolutist formulae that ignore geopolitical reality, gives realists pause. Unsurprisingly, political leaders who recognize the need to protect security interests find NPT Article VI a difficult provision to embrace—indeed, difficult to take seriously. Policy makers are prone to ask whether there is anything behind Article VI *other than* diplomats' sentiment.

We repeat familiar formulas, such as those that refer to the NPT as the cornerstone of nuclear non-proliferation. Repeating the formulas is good practice, perhaps, insofar as continuity and stability are valued in diplomatic settings.

However, to build a well-functioning nuclear weapons diplomacy that serves the interests of the US, allies, and like-minded countries, we must go beyond rote. If all that we do is extoll the NPT, then we do not speak to those who regard the NPT with well-earned skepticism. And we do nothing to advance arms control.

We need to use the NPT proactively, and to do that, we need to achieve the widest possible buy-in across the spectrum of mainstream opinion in our societies. Using the NPT to bring pressure to bear on China for its largely unaccountable nuclear weapons buildup is an important step. We should continue using the NPT in that way.

However, it is also time to revitalize our approach to negotiating under the NPT. To do *that*, we need fresh thinking about the substance that we might bring to the negotiating table. Forty years ago, observing a lack of motion on arms control between the US and USSR, Henry Kissinger wrote, 'The stalemate in negotiations reflects an impasse in thought.'[14] It is a central insight of realists that arms control and eventual disarmament are attainable only if we first address the substantive differences that divide us from our nuclear rivals. To do this, we must escape the 'impasse in thought' that bedevils arms control today. In this book, I have suggested a number of issues over which substantive differences exist and which are sufficiently concrete and practical that they might lend themselves to solutions through good faith negotiation. Here, too, a fresh approach to the NPT can bring the realists back to arms control. The stakes today are too high for us not to join the endeavor.

Notes

Introduction: A Tale of Two Treaties

[1] Proposals for international control began even before the USSR's first atomic weapon test, famously with the Acheson–Lilienthal Report of 1946.

[2] For historical context, see Willrich (1966) at 36–7.

[3] See Statute of the International Atomic Energy Agency (IAEA Statute) (1956).

[4] IAEA Statute, Art. II. As to which, see Freeman (1960) at 383–92.

[5] Treaty Banning Nuclear Weapons Tests in the Atmosphere, in Outer Space, and Under Water (1963).

[6] The President's News Conference, March 21, 1963 (107), reprinted in *Public Papers of the Presidents: John F. Kennedy, 1963*.

[7] Treaty on the Non-Proliferation of Nuclear Weapons (NPT) (1968).

[8] The declaration by North Korea (Democratic People's Republic of Korea—DPRK) on January 10, 2003 effectively terminated the DPRK's participation in the NPT but left unclear the DPRK's party status. The US counts 191 NPT member states: www.state.gov/nuclear-nonproliferation-treaty/

[9] Reference to the NPT as the 'cornerstone' of nuclear arms control is commonplace. See, for example, Working paper submitted by New Zealand on behalf of the New Agenda Coalition (Brazil, Egypt, Ireland, Mexico, New Zealand, and South Africa), March 15, 2018, Preparatory Committee (Prep. Com.) for 2020 Review Conference (Rev. Con.): NPT/CONF/2020/PC.11/WP.13, ¶4.

[10] In accordance with NPT Article VIII, paragraph 3, the NPT parties held a conference five years after the NPT's entry into force 'to review the operation of [the treaty] with a view to assuring that the purposes of the Preamble and the provisions of the Treaty are being realized.' At five-yearly intervals thereafter, further Review Conferences were convened on majority vote of the parties. In accordance with NPT Article X, paragraph 2, the NPT parties convened and decided (in 1995) that the Treaty shall continue in force indefinitely: 1995 Rev. Con. Decision 3, NPT/CONF.1995/32 (Part I) at 12 (May 12, 1995), https://documents-dds-ny.un.org/doc/UNDOC/GEN/N95/178/16/pdf/N9517816.pdf?OpenElement#page=17. The NPT parties also convene a Preparatory Committee ('Prep. Com.') in each of

the three years prior to a Rev. Con. See GAR 47/52 A, December 9, 1992.

[11] 'Nuclear-weapon State' is the expression that the NPT uses to refer to each state that 'has manufactured or exploded a nuclear weapon or other nuclear explosive device prior to 1 January, 1967': NPT Art. IX, para. 3. The nuclear-weapon states for purposes of the NPT therefore are China, France, Russia, the UK, and the US. Other states, for purposes of the NPT, are referred to as 'non-nuclear-weapon states.' In this book, I use the term 'state' as it is used in the NPT—that is to say, to refer to an independent sovereign country recognized by other states to have capacity to hold the full scope of rights and obligations available under international law.

[12] See, for example, Del Haggie, who has served as New Zealand's Ambassador for Disarmament and Permanent Representative to the Conference on Disarmament, in a discussion hosted by the Carnegie Endowment for International Peace in March 2017 said that 'the NPT is already endangered, already in peril.' Transcript of Carnegie International Nuclear Policy Conference 2017: www.apln.network/members/aotearoa-new-zealand (March 21, 2017). See also Susan F. Burk's remark, '[o]n the NPT in peril': *id.* at 24. Burk served as Special Representative of the President for Nuclear Nonproliferation (US) 2009 to 2012.

[13] As to part of this bargain, see Bourantonis (1997) at 357.

[14] Similarly, states that stated that Art. VI convinced them to become party to the NPT were making a political choice and stating a reason for doing so. They were not making a treaty reservation. See, for example, Working Paper (Ghana, Mexico, Morocco, Nigeria, Peru, Romania, Sudan, Yugoslavia, and Zaire), May 12, 1975, NPT/Conf/18 p 2. As French foreign ministry advisors noted in 1968, the NPT was a 'project, the intention, *if not the content*, of which meets the wishes of the vast majority of non-nuclear weapon States' (emphasis added): Note on Guarantees for non-nuclear-weapon States (Directorate of Political Affairs, Ministry of Foreign Affairs, France) chapeaux (March 29, 1968), reprinted Wilson Center Digital Archive.

[15] NPT/CONF.1995/32 (Part I), Annex.

[16] Decision 3's 'emphasizing' Decision 2—Principles and Objectives for Nuclear Non-Proliferation and Disarmament (NPT/CONF.1995/32 (Part I), Annex)—does not add a conditional element to the NPT either. The undiluted primacy of non-proliferation in the treaty remains evident in paragraph 2 of Decision 2, which affirms the 'vital role' of the NPT in preventing nuclear proliferation and contains no conditionality or linkage to disarmament.

[17] Treaty on the Prohibition of Nuclear Weapons (TPNW) (2017).

[18] TPNW Art. 4(1).

[19] With credit for the paraphrase to Charles Dickens and the late James Crawford, Australian Judge of the International Court of Justice: Crawford (2017) at 459.

[20] See further Grant, *China's Nuclear Build-Up* (2021) at 40; Grant (2021–22) at 28–44; and this book, Chapter 4.

one Three Pillars or One Foundation?

[1] See, for example, the Hon. Christopher A. Ford, Assistant Secretary for International Security and Nonproliferation, Opening Statement by the United States of America, Prep. Com. for the 2020 Review Conference of the Parties to the NPT (April 23, 2018), Geneva at 5: http://statements.unmeetings.org/media2/18559133/usa-2018-npt-prepcom_us-general-debate-statement-copy.pdf#page=5

[2] *Id.*

[3] See GAR 2373 (XXII), June 12, 1968, ¶3.

[4] Nye described the NPT as doing precisely that—limiting the otherwise sovereign right to self-defense. See Nye (1985) at 125.

[5] Regarding 'mutual nonacquisition pacts' embodied in the NPT, see Knopf (2012/13) at 95.

[6] Ford in Lavoy and Wirtz (eds.) (2012) at 184.

[7] See to this effect Zarate in Sokolski (ed.) (2010) at 222.

[8] 729 UNTS at 172.

[9] See IAEA INFCIRC/153, ¶¶80, 81, 106.

[10] Ford in Lavoy and Wirtz (eds.) (2012) at 185.

[11] www.iaea.org/sites/default/files/publications/documents/infcircs/1970/infcirc140.pdf

[12] www.iaea.org/sites/default/files/publications/documents/infcircs/1972/infcirc153.pdf

[13] Prior to INFCIRC/153, the IAEA concluded safeguards agreements piecemeal under INFCIRC/66. The Board of Governors had approved INFCIRC/66 in 1965 and provisionally extended it in 1966 and 1968. See further Willrich (1966) at 38–48.

[14] Regarding Iraq, see Kelley (2023) at 12–22.

[15] www.iaea.org/sites/default/files/infcirc540c.pdf. As to the evolving safeguards system, see Hooper (2003) at 11.

[16] www.iaea.org/topics/safeguards-agreements. For a list, see www.iaea.org/sites/default/files/20/01/sg-agreements-comprehensive-status.pdf

[17] www.iaea.org/sites/default/files/20/01/sg-ap-status.pdf

[18] See Bellamy (2005) at 76–103. See also Joyner (2011) at 85.

[19] NPT Art. III, para. 1.

[20] In contrast to a safeguards agreement, it is not legally required that an NPT state conclude an Additional Protocol—that is to say, a protocol modeled on INFCIRC/540.

[21] SCR 1887 (2009), September 24, 2009, ¶15.

[22] See further Suseanu (2021).

[23] The Marshall Islands thought the concept of a three-pillars 'regime' came from the Final Document of the 1985 NPT Review Conference: *Marshall Islands Memorial* (March 16, 2015) at 73 n. 266.

[24] See, for example, Mallard (2014) at 449.

[25] See www.state.gov/nuclear-nonproliferation-treaty/

[26] See Lord Ahmad of Wimbledon, Minister of State, Foreign, Commonwealth and Development Office (UK), HoL Hansard (UK), vol. 817 col. 1082 (January 12, 2022).

[27] Joyner (2011) at 75 (emphasis original).

[28] Sanders (2004).

[29] Bolton (2004).

[30] Consider Joyner's *Lotus* argument about the freedom of non-nuclear-weapon states: Joyner (2011) at 83. The *Lotus* principle, so-called for the role it played in *The Case of the 'S.S. Lotus' (France v. Turkey)* (1927), in its 'classical formulation' holds that 'whatever is not explicitly prohibited by international law is permitted': see Hertogen (2016) at 902 with works cited *id.* at 902 n. 6.

[31] See Mexico's statement, quoted by Ford (2010) at 313.

[32] See Note No. 307/SJ RJ, ¶II (March 18, 1968), reprinted in Wilson Center Digital Archive.

[33] Jonas and Braunstein (2018) at 351–75.

[34] ElBaradei (2006).

[35] Albert Wohlstetter, Greg Jones, and Roberta Wohlstetter, *Towards a New Consensus on Nuclear Technology*, Vol. 1, 1979, PH 78-04-832-33, a summary report prepared for the US Arms Control and Disarmament Agency, Contract #AC7 NC-106 (Los Angeles, CA: PanHeuristics, July 6, 1979) 34–5, cited at Zarate (2010) at 230 n. 7 and by Ford (2010) at 237.

[36] Ford (2010) at 238.

[37] *Id.* at 335.

[38] It was recognized at the time of the NPT's adoption that non-aligned states 'have an advantage in seeing their neighbors make the commitment not to acquire [nuclear weapons]': Note on the Question of the Non-Proliferation of Nuclear Weapons (Directorate of Political Affairs, Ministry of Foreign Affairs, France) (April 3, 1968) at 17, reprinted in Wilson Center Digital Archive. See also, generally, Knopf (2012/2013).

[39] An 'elastic' theory of the NPT's core provisions finds reflection in practice. Brazil, for example, in refusing to adopt an IAEA Additional Protocol, refers to Art. VI and nuclear-weapon states' slowness in pursuing disarmament negotiations: Valle Machado da Silva (2021) at 386–7. See also comparative observations regarding Argentina (which, in tandem with Brazil, refuses to adopt an Additional Protocol): Pretorius (2013) at 396–7. Cf. Bezerra (2021) at 39–49.

[40] Judge Anzilotti thought the principle *inadimplenti non est adimplendum* belongs to general international law and that, invoking it, a party might be excused the non-observance of certain treaty terms, when another party failed to observe related terms. See *Diversion of Water from the Meuse*, June 28, 1937, Dissenting Opinion, Judge Anzilotti, PCIJ Rep. ser. A/B No. 70 at 50. However, parties have found courts and tribunals reluctant to accept pleas under that principle. See, for example, Greece's pleadings against the former Yugoslav Republic of Macedonia (as North Macedonia then was called) in *Application of the Interim Accord of September 13, 1995 (the former Yugoslav Republic of Macedonia v. Greece)*, 2011 ICJ at 644 (December 5). (N.B.: I served as Counsel to Greece in that case.)

[41] *Gabčíkovo-Nagymaros Project (Hungary v. Slovakia)* (1997). See also *Temple of Preah Vihear (Cambodia v. Thailand)* (1962) at 57 (separate opinion, Fitzmaurice, J.).

[42] See VCLT Art. 60.

[43] India and Sweden, for example. See Note on the Question of the Non-Proliferation of Nuclear Weapons (Directorate of Political Affairs, Ministry of Foreign Affairs, France) (April 3, 1968) at 13, reprinted in Wilson Center Digital Archive. Further to India and Sweden on this point, see Harries (May 2015) at 4–5; and as to Burma, Nigeria, and the United Arab Republic as well, Ford (2007) at 405–6.

[44] In a 2015 paper, Matthew Harries reaches a similar conclusion and posits that 'it is reasonable for non-nuclear weapon states to regard the commitment to progress on disarmament as a fundamental element of NPT *politics*' (emphasis added). Harries (2015) at 4.

[45] See, for example, Kirsten and Zarka (2022) at 4; Knopf (2012/13) at 112.

[46] 729 UNTS at 173.

[47] See Grant, *China's Nuclear Build-Up* (2021) at 46; Grant (2021–22) at 65.

[48] See Floyd D. Spence National Defense Authorization Act for Fiscal Year 2001, sec. 1041 (Revised Nuclear Posture Review).

[49] https://media.defense.gov/2022/Mar/29/2002965339/-1/-1/1/FACT-SHEET-2022-NUCLEAR-POSTURE-REVIEW-AND-MISSILE-DEFENSE-REVIEW.PDF

[50] *2022 National Defense Strategy of the United States of America including the 2022 Nuclear Posture Review and the 2022 Missile Defense Review* (NPR).

51 2022 NPR at 17.

52 *Id.* at 18.

53 Nye (1985) at 125.

54 *Legality of the Threat or Use of Nuclear Weapons,* Advisory Opinion (1996) at 264, ¶99.

55 See Grant, *China's Nuclear Build-Up* (2021), at 11–12; Grant (2021–22) at 12–16.

56 GAR 49/75 K, December 15, 1994, asked the ICJ for its advisory opinion on the question, 'Is the threat or use of nuclear weapons in any circumstance permitted under international law?' See the trenchant critique at the time by Dapo Akande (now Chichele Professor of International Law, Oxford University and Member, UN International Law Commission): Akande (1998) at 196.

57 www.westminster.ac.uk/news/professor-marco-roscini-appointed-as-swiss-chair-of-international-humanitarian-law

58 Roscini (2015). A cautionary word is in order about Roscini's use of the word 'vague' in describing Article VI, a description that other writers have used as well; for example, Müller and Wunderlich (2020) at 174. There are treaty provisions, such as NPT Article VI, that leave discretion in the hands of the states to which they apply; discretion, however, is not necessarily vagueness. The better view is that NPT Article VI is clear in what it requires—it requires the parties to pursue negotiations—and leaves in the hands of the parties the discretion how to implement the requirement.

59 NPT/CONF.1995/32 (Part I), Annex.

60 *Id.*

61 Costlow, 'Review' (2022) at 148–9.

two The TPNW Challenge

1 Protocol for the Prohibition of the Use in War of Asphyxiating, Poisonous or Other Gases, and of Bacteriological Methods of Warfare (1925).

2 Convention on the Prohibition of the Development, Production and Stockpiling of Bacteriological (Biological) and Toxin Weapons and their Destruction (1972).

3 Convention on the Prohibition of the Development, Production, Stockpiling and Use of Chemical Weapons and on their Destruction (1993).

4 Convention on the Prohibition of the Use, Stockpiling, Production and Transfer of Anti-Personnel Mines and on their Destruction (1997).

5 Convention on Cluster Munitions (2008).

6 For reasoned objections to US participation in the Landmine Convention, see www.heritage.org/global-politics/report/the-ottawa-mine-ban-convention-unacceptable-substance-and-process (December 13, 2010).

[7] *Legality of the Use by a State of Nuclear Weapons in Armed Conflict*, Request for Advisory Opinion, World Health Assembly resolution of May 14, 1993; *Legality of the Threat or Use of Nuclear Weapons*, Request for Advisory Opinion, GAR 49/75K, December 15, 1994.

[8] The first resolution of the first session of the UN General Assembly in fact called for, *inter alia*, 'the elimination from national armaments of atomic weapons and of all other major weapons adaptable to mass destruction': GAR 1(I), January 24, 1946, ¶5*(c)*. Consider as well civil society endeavors; for example, the Federation of Atomic Scientists (later re-titled Federation of American Scientists), founded in 1945: https://fas.org/about-fas/

[9] SG/SM/12484DC/3192 (September 24, 2009): https://press.un.org/en/2009/sgsm12484.doc.htm

[10] Regarding the emergence of the TPNW and with links to documentation, see United Nations Office for Disarmament Affairs, *Treaty on the Prohibition of Nuclear Weapons*, available at www.un.org/disarmament/wmd/nuclear/tpnw/

[11] The qualifier 'most' is justified, because scenarios might arise in which the immediate consequences of the use of a nuclear weapon would *not* entail a breach of humanitarian law. To give perhaps the main example, there is the use of nuclear weapons at sea against naval targets in isolation from civilians. See Mizokami (2020). The US understands that the use of a nuclear weapon is 'not inherently disproportionate': *Law of War Manual* (December 2016) at 367, ¶6.7.4. The *Law of War Manual* thus identifies use cases in accordance with the law of armed conflict.

[12] 2010 Review Conference Final Document, vol. I, at 12, ¶80: NPT/CONF.2010/50, available at https://undocs.org/NPT/CONF.2010/50%20(VOL.I)

[13] Noting the link between the 2010 Review Conference, the Final Document, and the transactions in the ban conference that the General Assembly eventually called, see Arimatsu (2021) at 892. As to the role played in promoting the TPNW by the International Campaign to Abolish Nuclear Weapons (ICAN), a non-governmental organization that received the Nobel Prize in 2017 for its activities in that direction, see Slatter (2020) at 6.

[14] UN background document on the TPNW and its drafting history, available at www.un.org/disarmament/tpnw/background.html (emphasis added).

[15] *Id.*

[16] Documents such as the US Department of Defense Law of War Manual (as updated July 2023) evince the concern, not only as statements available to a public audience for scrutiny and debate but

also as operational directions that control and limit the conduct of the armed forces.

[17] 2010 Review Conference Final Document vol. I at 20 (under Conclusions and Recommendations for follow-on actions, item B(*i*)).

[18] See Arms Control Association (2022).

[19] 729 UNTS at 173. As to which see US Assistant Secretary of State for Arms Control Stephen Rademaker (2005): '[T]he language [of the NPT] contains no suggestion that nuclear disarmament is to be achieved before general and complete disarmament.'

[20] Writers indeed have noted that participation in a nuclear weapons ban treaty would not accord with participation in defensive alliances, including NATO and ANZUS, which rely upon nuclear deterrence. See, for example, Hood and Cormier (2020) at 161: 'it is unlikely to be possible for states to simply ignore existing security frameworks.'

[21] GAR 71/258 (2016), December 23, 2016.

[22] See UN Office for Disarmament Affairs, *Treaty on the Non-Proliferation of Nuclear Weapons (NPT)*, www.un.org/disarmament/wmd/nuclear/npt/

[23] A/CONF.229/2017/8 (July 7, 2017).

[24] In accordance with TPNW Art. 15.

[25] See www.nti.org/education-center/treaties-and-regimes/treaty-on-the-prohibition-of-nuclear-weapons/

[26] For the list of parties having signed and ratified the TPNW, see https://treaties.unoda.org/t/tpnw/participants.

[27] UNTS No. 56487 at 38.

[28] *Id.*

[29] *Id.* at 39–40.

[30] *Id.*

[31] Kristensen and Korda (2021).

[32] Under this approach, it would appear that more than one 'plan' for weapons elimination would emerge. As to the problem that the emergence of multiple plans would present, see former US State Department lawyers Highsmith and Stewart (2018) at 135.

[33] See the trenchant critique by Highsmith and Stewart (2018) at 147 n. 9.

[34] When critiquing the TPNW, the US has recalled the history of US–Russia arms control. See, for example, Mr. Wood (United States of America), A/C.1/72/PV.12 p 7 (October 12, 2017).

[35] A salient example is the EU's treaty-making project to replace investment arbitration with an international investment court. See, for example, EU–Mexico draft text, Art. 14, calling on the parties to cooperate 'for the establishment of a multilateral mechanism for the resolution of investment disputes,' Modernisation of Trade part of the EU–Mexico

Global Agreement, at 9–13, available at https://trade.ec.europa.eu/doclib/docs/2018/april/tradoc_156814.pdf. The EU–Viet Nam Investment Protection Agreement provides that the parties 'shall enter into negotiations for an international agreement providing for a multilateral investment tribunal in combination with, or separate from, a multilateral appellate mechanism,' EU–Viet Nam Investment Protection Agreement (June 30, 2019), Art. 3.41, available at https://investmentpolicy.unctad.org/international-investment-agreements/treaty-files/5868/download

[36] Disarmament activists suggest that the prohibition in the TPNW reflects a rule of customary international law and thus binds all states. However, the prohibition is a treaty rule and thus limited in application to the TPNW parties. See Grant (2022) at 1–54.

[37] Slatter (2020) at 6–7.

[38] Kimball (2020).

[39] Koplow and Ifft (2021).

[40] Müller and Wunderlich (2020), Abstract, at 171.

[41] Graham (2020) at 234.

[42] UnfoldZero says that the TPNW 'supports the NPT and serves as part of NPT implementation by non-nuclear States': www.unfoldzero.org/treaty-on-the-prohibition-of-nuclear-weapons/. SIPRI says that 'the core obligations of the two treaties … seem to be perfectly consistent with one another' though acknowledges that there is a 'perceived incompatibility between the two' having 'mainly … to do with the indirect negative consequences that the TPNW could potentially have on the NPT's non-proliferation objectives.' Erästö (2019). See also Müller and Wunderlich (2020) at 172: 'Reasonable policies can create a *modus vivendi* [between the NPT and TPNW].'

[43] See, for example, Working paper submitted by New Zealand on behalf of the New Agenda Coalition, March 15, 2018, Prep. Com. for 2020 Rev. Con.: NPT/CONF/2020/PC.11/WP.13, ¶4.

[44] Müller and Wunderlich (2020) at 172.

[45] *Id.* at 184.

[46] For the foremost rebuttal of the critical theory of 'radical indeterminacy' in international law (and a definition), see Crawford (2013) at 113–35.

[47] Müller and Wunderlich (2020) at 172.

[48] For a psychologizing assessment of the disarmament activists' faded hopes and a resultant search for 'emotional satisfaction,' see Müller and Wunderlich (2020) at 177.

[49] See, in particular, *Obligations concerning Negotiations relating to Cessation of the Nuclear Arms Race and to Nuclear Disarmament (Marshall Islands v. United Kingdom)* (2016). The Marshall Islands asked the ICJ to hold that the UK had 'violated and continues to violate its international obligations

under the NPT, more specifically under Article VI of the Treaty, by failing to pursue in good faith and bring to a conclusion negotiations leading to nuclear disarmament,' Judgment, 2016 ICJ at 833, 839, ¶11. The ICJ declined to reach the merits of the Marshall Islands' request, on the ground that the Marshall Islands had not identified a dispute with the UK subject to the ICJ's jurisdiction: ICJ 2016 at 856, ¶58.

[50] See also the slightly different observation, by US Secretary of Defense Austin at the Halifax International Security Forum, that Russia's aggression gives a proliferation incentive to would-be aggressor-states, because they 'could well conclude that getting nuclear weapons would give them a hunting license of their own.' US Department of Defense, 'Austin Says Ukraine Matters to all Peoples and Nations' (November 19, 2022): www.defense.gov/News/News-Stories/Article/Article/3224096/austin-says-ukraine-matters-to-all-peoples-and-nations/. As to continuity on this point between the Trump Administration and the Biden Administration, with intellectual groundwork found earlier still, see Ford (2023) at 9.

[51] See *United Nations Disarmament Yearbook 2019: Part II*, at 246. The chart in the *Yearbook* might suggest that the TPNW soon will overtake the CWC, NPT, BWC, and CTBT in number of participants, but a cautious reader will see that the chart compares unlike things: it shows the rate at which new participants joined the TPNW in the *first two years* after the TPNW opened for signature and compares that curve with one depicting the rate at which new participants had joined the four older arms control treaties *decades after* they had opened for signature. The rhetorical effect is to present the TPNW as a treaty on the move, the others at a plateau, but in truth the contrast is probably not so great.

[52] See Nuclear Weapons Ban Monitor, 'The obligation to promote universal adherence to the Treaty', https://banmonitor.org/positive-obligations-1/the-obligation-to-promote-universality-of-the-treaty

[53] https://treaties.unoda.org/t/tpnw

[54] 2022 NPR at 19.

[55] Adelman observed in particular that civil society in democratic countries places pressure on governments to comply with arms control agreements; society in authoritarian countries does not. Adelman (1984) at 245.

[56] See Schelling (1960) (asymmetry in impact of threats); Schelling and Halperin (1961) at 54 (asymmetry in impact of arms reductions); Bellamy (2005) at 170 (asymmetry in arms control verification); Kroenig (2010) at 173 (asymmetry in proliferation risk). See also Adelman (1984) at 243.

[57] Müller and Wunderlich (2020), Abstract at 171; and further developed *id.* at 181.

[58] Heft (1991) at 165.

[59] *Id.* at 172.

[60] *Id.* at 230.
[61] *Id.* at 60–1.
[62] Evangelista (2010) at 414–15, 417.
[63] Van Cleave in Staar (ed.) (1984) at 3 (emphasis original).
[64] P5 Joint Statement on the TPNW (2018). As to the non-effect of the TPNW on customary international law, see Grant (2022) at 1–53.
[65] China, Implementation of the Treaty on the Non-Proliferation of Nuclear Weapons (2019) at ¶31.
[66] *Id.* at ¶79.
[67] The Vienna Group of Ten has convened for NPT Prep. Com. and Rev. Con. meetings since 1980 to advance agreed non-proliferation issues that they refer to as 'Vienna issues.' See Vienna Group of Ten, May 2017 NPT Prep. Com., General Debate Statement: https://geneva.mission.gov.au/gene/Statement776.html
[68] NPT/CONF.2020/PC.III/WP.2 (March 15, 2017), Addressing 'Vienna issues,' Working paper submitted by Australia, Austria, Canada, Denmark, Finland, Hungary, Ireland, the Netherlands, New Zealand, Norway, and Sweden (the Vienna Group of Ten), ¶1.
[69] See, for example, GAR 75/40, December 7, 2020, and for voting record, A/75/PV.37 (December 7, 2020) at 15.
[70] For the October 28, 2022 GAR in draft, see https://reachingcriticalwill.org/images/documents/Disarmament-fora/1com/1com22/resolutions/L17.pdf
[71] See, for example, 'States reaffirm support for Nuclear Weapons Ban Treaty at UN First Committee,' *ICAN Updates* (October 29, 2022): www.icanw.org/states_reaffirm_support_for_nuclear_weapons_ban_treaty_at_un_first_committee
[72] *Labour's Manifesto* (June 2024), https://labour.org.uk/change/strong-foundations/
[73] See Francis (2024). See also 'MPs vote to renew Trident weapons system,' *BBC News* (July 19, 2016), www.bbc.co.uk/news/uk-politics-36830923. Forty-seven Labour members at the time voted against renewing the Trident system. *Id.* For the Commons debate and vote on Trident (472 ayes, 117 noes), see HC Hansard, vol. 613, cols 656–9 (July 18, 2016).
[74] See HoL debate of January 21, 2012 (that is, the day before the entry into force of the TPNW); for example, question of The Lord Bishop of Coventry: HoL Hansard, vol. 809, col. 1268 (January 12, 2021).
[75] See Baroness Miller of Chilthorne Domer, HoL Hansard, vol. 809, col. 1270 (January 21, 2021); and, generally, ICAN (2020).
[76] See Final Document of the 18th Non-Aligned Movement Summit of Heads of State and Government (2019) at 81, ¶245: the Non-Aligned Movement Heads of State and Government '*took note* of the adoption

of the [TPNW] …' which, they 'hoped' will 'contribute to furthering the objective of the total elimination of nuclear weapons.'

77 Akita (2022).

78 Laikola and Rolander (2022); International Institute for Strategic Studies (2023).

three Article VI Interpreted and Applied

1 For drafting history of Art. VI, see Ford (2007) at 403; and chapters in Popp, Horovitz, and Wenger (eds.) (2017), *passim*. The Arms Control Association supplies a timeline: www.armscontrol.org/factsheets/Timeline-of-the-Treaty-on-the-Non-Proliferation-of-Nuclear-Weapons-NPT. For the procedural history, with links to documents including First Committee summary records and relevant General Assembly resolutions, see UN Audiovisual Library of International Law, Treaty on the Non-Proliferation of Nuclear Weapons: https://legal.un.org/avl/ha/tnpt/tnpt.html

2 See Chapter 1.

3 For my earlier treatment, see Grant (2021–22) at 8–12, 27–47.

4 See, later this chapter, public remarks of US Amb. Jackie Sanders and US Deputy Legal Advisor Ronald Bettauer.

5 See Grant (2021–22) at 27–47. See also Grant, *China's Nuclear Build-Up* (2021) at 40.

6 See esp. Trachtenberg, Dodge, and Payne (2021), *passim* and at 11–20.

7 Testimony of Secretary of Defense Harold Brown before the US Congress, House of Representatives, Committee on the Budget, Outlook and Budget Levels for Fiscal Years 1979 and 1980: Hearings Before the United States House of Representatives Committee on the Budget, 96th Congress, 1st Session, Part 1 (Washington, DC: US Government Printing Office, 1979), at 500, quoted in Trachtenberg et al (2021) at 3, ¶7.

8 As second nature as the NPT's categories might seem today, the categories drew scrutiny and critique at the time of the NPT's adoption. See, for example, Note on Guarantees for non-nuclear-weapon States (Directorate of Political Affairs, Ministry of Foreign Affairs, France) at ¶2 ('Discrimination between States') (March 29, 1968), reprinted in Wilson Center Digital Archive. Cf. NATO Council meeting of February 22, 1967, reprinted in *Negotiations of Articles I and II of the NPT. Selected Documents*, vol. 2 (1966–68) at 9–10.

9 Bettauer (2006).

10 Sanders (2005).

[11] Even a cabinet-level official expressly declaring a change in a country's legal position under a treaty does not necessarily effectuate the change for purposes of international law. See *Armed activities on the territory of the Congo (New Application: 2002) (Democratic Republic of Congo v. Rwanda)*, Judgment, 2006 ICJ 6, at 27–9, ¶¶46–55.

[12] The NPT is a multilateral treaty; its terms do not change through the conduct of just one party. See *Guiding Principles Applicable to Unilateral Declarations of States Capable of Creating Legal Obligations* (2006) at 163 Guiding Principle 4 Comment 1.

[13] For my earlier treatment, see Grant (2021–22) at 12–17.

[14] See *Maritime Delimitation in the Indian Ocean (Somalia v. Kenya)*, Preliminary Objections, Judgment, 2017 ICJ at 3, ¶90 (February 2); *Land and Maritime Boundary between Cameroon and Nigeria (Cameroon v. Nigeria: Equatorial Guinea intervening)*, Judgment, 2002 ICJ at 303, ¶244 (October 10).

[15] *Somalia v. Kenya*, 2017 ICJ at 3, 37, ¶90.

[16] *Somalia v. Kenya*, Joint Declaration, Judges Gaja and Crawford, 2017 ICJ at 63, ¶4 (emphasis original).

[17] A prominent example is Russia in claims that Georgia and Ukraine instituted against it at the ICJ. See *Application of the Int'l Convention on the Elimination of All Forms of Racial Discrimination (Georgia v. Russian Federation)*, Preliminary Objections, Judgment, 2011 ICJ at 70, 120, ¶116 (April 1); *Application of the Int'l Convention for the Suppression of the Financing of Terrorism and of the Int'l Convention on the Elimination of All Forms of Racial Discrimination (Ukraine v. Russian Federation)*, Preliminary Objections, Judgment, 2019 ICJ at 558, ¶¶66–70 (November 8). Cf. pre-arbitration waiting periods stipulated in some treaties, such as that in Art. 26(2) of the Energy Charter Treaty, and invoked from time to time against treaty claimants: see, for example, *Stati et al v. Kazakhstan, SCC Arb V (116/2010) (Böckstiegel, Chairman; Haigh & Lebedev, Co-Arbitrators)*, Award (December 19, 2013), at 180–2, ¶¶820–7. (N.B.: I served as legal expert for Kazakhstan in post-award national court proceedings in the *Stati* case.)

[18] *Application of the Int'l Convention for the Suppression of the Financing of Terrorism and of the Int'l Convention on the Elimination of All Forms of Racial Discrimination (Ukraine v. Russian Federation)*, 2019 ICJ at 588, ¶69. See also *Appeal relating to the Jurisdiction of the ICAO Council (Bahrain, Egypt and United Arab Emirates v. Qatar)*, Judgment, 2020 ICJ at 172, 201, ¶90 (July 14); *Continental Shelf (Tunisia/Libya)* 1982 ICJ, at 145, ¶4 (dissenting opinion by Gros, J.), citing, *inter alia*, *Lake Lanoux* Award, XII UNRIAA 281, 311 (November 16). *Cf.* Abdulqawi Ahmed Yusuf, 'Engaging with International Law' (2020) 69 *ICLQ* at 505, 515.

[19] *Application of the Interim Accord of September 13, 1995 (the former Yugoslav Rep. of Macedonia v. Greece)*, 2011 ICJ at 685, ¶134 (emphasis added) (December 5).

[20] *Georgia v. Russian Federation*, 2011 ICJ at 132, ¶157.

[21] The Macedonian party at the time was titled for purposes of the Interim Accord, and for transactions in the UN, as 'the Former Yugoslav Republic of Macedonia' (the FYROM) in accordance with SCR 817 (1993) and SCR 845 (1993). The name 'Republic of North Macedonia,' short form 'North Macedonia,' was agreed between the two states under Article 1(3)(a) of the Final Agreement for the settlement of the difference described in the United Nations Security Council Resolutions 817 (1993) and 845 (1993) (Prespa Agreement) (2018).

[22] Interim Accord, Art. 5(1), October 13, 1995, 1891 UNTS at 3, 5.

[23] *The former Yugoslav Republic of Macedonia v. Greece*, 2011 ICJ at 692, ¶166. For some earlier examples of such negotiation clauses—that is, clauses requiring negotiation but not as a pre-condition for recourse to a court or arbitral tribunal—*see* Rogoff (1994) at 161–71.

[24] 2011 ICJ at 686, ¶135.

[25] Proceedings in the *North Sea Continental Shelf* cases were instituted on February 20, 1967. See the Special Agreements (Denmark-Germany; Germany-Netherlands) submitting the matter to the ICJ: *North Sea Continental Shelf*, Pleadings, Oral Arguments, Documents, 1968 ICJ 5, at 6–9. As of December 7, 1967, 'substantial progress' had been made on the draft NPT text, but 'a final draft [had] not as yet been achieved': Interim Report of the Conference of the Eighteen-Nation Committee on Disarmament, A/6951, at 2, ¶5 (December 7, 1967).

[26] *North Sea Continental Shelf (Germany/Denmark; Germany/Netherlands)*, Judgment, 1969 ICJ at 3, ¶85 (February 20).

[27] See, for example, *The former Yugoslav Republic of Macedonia v. Greece*, 2011 ICJ at 684, ¶131. See also *Somalia v. Kenya*, Joint Declaration of Judges Gaja and Crawford, 2017 ICJ at 63–4, ¶5. For an example of a national statute that does stipulate good faith in the negotiation setting, see the Native Title Act 1993 (Cth) as applied in *Charles v. Sheffield Resources Ltd* [2017] FCAFC at 218, ¶3 (Federal Ct. Austral.) (noting the 'correlative obligation on other persons to negotiate in good faith').

[28] *Border and Transborder Armed Actions (Nicaragua v. Honduras)*, Jurisdiction and Admissibility, 1988 ICJ at 69, ¶94 (December 20) (internal quotations omitted), quoting *Nuclear Tests cases (Australia v. France; New Zealand v. France)*, Judgment, 1974 ICJ at 253, ¶46; at 457, ¶49 (December 20).

[29] '[T]he General Assembly can hardly be supposed . . . to ask the [ICJ's] opinion as to the reasons which, in the mind of a Member, may prompt its vote.' *Conditions of Admission of a State to Membership in the United*

Nations (Article 4 of the Charter), Advisory Opinion, 1948 ICJ at 57, 60 (May 28).

[30] As to factors relevant to determining whether a state intends to recognize another entity as a state, see Grant in Chinkin and Baetens (eds.) (2015) at 192.

[31] *Somalia v. Kenya*, 2017 ICJ at 64, ¶5.

[32] *Nicaragua v. Honduras*, 1988 ICJ at 105, ¶94.

[33] *Cf. NLRB v. Katz*, at 369 U.S. at 736, 743 (Brennan, J., 1962): 'Clearly, the duty [to negotiate] may be violated without a general failure of subjective good faith; for there is no occasion to consider the issue of good faith if a party has refused even to negotiate in fact—"to meet . . . and confer"—about any of the mandatory subjects.'

[34] *Georgia v. Russian Federation*, 2011 ICJ at 133, ¶161.

[35] *Id.* at 139, ¶180: '[R]egardless of the Russian Federation's ambiguous and perhaps conflicting statements on the subject of negotiations with Georgia as a whole, and President Saakashvili personally, these negotiations did not pertain to CERD-related matters.'

[36] *Id.*

[37] The stepwise approach was visible from the start. GAR 2028 (XX), November 19, 1965, ¶ 1*(c)* envisaged the eventual non-proliferation treaty as 'a step towards the achievement of general and complete disarmament and, more particularly, nuclear disarmament.' As to the relevance of a treaty preamble in interpreting an operative provision of the treaty, this follows from the character of the preamble as part of the context of the treaty, as recognized in the chapeau to Art. 31(2) VCLT: 'The context for the purpose of the interpretation of a treaty shall comprise, in addition to the text, ... its preamble and annexes,' VCLT Art. 31(2). Under the NPT, there is also the operative provision calling on the states parties to 'assur[e] that the purposes of the Preamble ... are being realised': NPT Art. VIII, para. 3.

[38] 2022 NPR at 16. This was in connection with Russia's introduction of new strategic weapons, which the 2022 NPR notes impedes New Start Treaty negotiations.

[39] Remarks by US Assistant Secretary of State for Arms Control Rademaker (2005). Some arms control academics and non-nuclear-weapon states' officials question this position. See, for example, Loose (2007) at 139–40, who posits that '[d]e-linking nuclear disarmament from general and complete disarmament' is necessary to 'the nuclear disarmament agenda.'

[40] Regarding the 'effective measures' head of Art. VI, see Dunworth (2016) at 601; Cormier and Hood (2017) at 27–46.

[41] Xinjun (2009) at 36. Xinjun was an associate professor at the School of Law of Tsinghua University, Beijing, at the time and as of December 2020, *inter alia*, an editor of the *Chinese Journal of International Law*.

42 *See* Ford (2007) at 403.

43 Academic writers who doubt Ford's limitive interpretation of the 'to pursue negotiations' clause have said that it is not a good faith interpretation. That overstates the matter. See, for example, Cormier and Hood (2017) at 35 and other works cited *id.* at 35 n. 138.

44 *Legality of the Threat or Use of Nuclear Weapons*, Advisory Opinion, 1996 ICJ at 66, 263, ¶99 (July 8) (emphasis added). Contrary to the ICJ's assertion, there is little evidence that general international law before the Advisory Opinion had contained a disarmament rule, or contains one today.

45 *Obligations concerning Negotiations relating to Cessation of the Nuclear Arms Race and to Nuclear Disarmament (Marshall Islands v. UK)*, Judgment, 2016 ICJ at 844, ¶20 (October 5); *Obligations concerning Negotiations relating to Cessation of the Nuclear Arms Race and to Nuclear Disarmament (Marshall Islands v. India)*, 2016 ICJ 255 at 264, ¶19 (October 5); *Obligations concerning Negotiations relating to Cessation of the Nuclear Arms Race and to Nuclear Disarmament (Marshall Islands v. Pakistan)*, 2016 ICJ 552 at 561, ¶19 (October 5).

46 See 1996 ICJ at 311–29. See also Wright (2009) at 228–9. Wright acknowledged the critics but begged to differ.

47 729 UNTS at 264.

48 *Id.* at 273.

49 *Railway Traffic between Lithuania and Poland*, Advisory Opinion, 1931 PCIJ Rep. Ser. A/B No. 42, at 116 (October 15).

50 United Nations Convention on the Law of the Sea (UNCLOS), Art. 283.

51 UNCLOS was not the first treaty to distinguish between 'exchange of views' and other forms of engagement. For example, air transport agreements of the Federal Republic of Germany concluded before UNCLOS (and before the NPT) have distinguished between 'exchange of views' and a consultation process. See Air Transport Agreement Art. 12–13, July 22, 1959, 464 UNTS at 189, 201–3; Air Transport Agreement Art. 8–9, Ger./Sen., October 29, 1964, 728 UNTS at 121, 125.

52 *Oxford English Dictionary*, 3rd edn, 2007.

53 For examples, see Grant (2021–22) at 22–3 n. 82.

54 See Rule 34 of the Rules of Court of the European Court of Human Rights, as amended December 13, 2004, and September 19, 2016; and European Convention on Human Rights, Art. 59.

55 *Andronicou and Constantinou v. Cyprus*, Case 86/1996/705/897, Judgment of October 9, 1996, at ¶165.

56 However, 'the only authentic version of the judgment . . . is that which appears in the language of the case': Language Arrangements, Court of

Justice of the European Union, https://curia.europa.eu/jcms/jcms/Jo2_10739/en/. As to translation functions in the Court of Justice, see Legal Translation at the Court of Justice of the European Communities, EULITA, https://eulita.eu/legal-translation-court-justice-european-communities/

[57] *Case T-271/04, Citymo SA v. Commission*, Judgment of the Court of First Instance (Second Chamber), [2007] ECR II-01375 (May 8, 2008).

[58] The US Court of Appeals for the Ninth Circuit thus cogently observed that 'the essential details of the negotiations [stipulated under Art. VI]—their time, their place, their nature—was unspecified upon ratification.' *Marshall Islands v. United States*, 865 F.3d at 1195. See also Matheson (1997) at 434: 'the [ICJ's advisory opinion] does not dictate any timetable or negotiating forum.'

[59] By contrast to provisions telling parties simply to pursue negotiations, certain provisions indicate a timeline or procedure. For example, NPT Article III, paragraph 4, requires that non-nuclear-weapon states parties commence negotiation of safeguard agreements 'within 180 days from the original entry into force' of the NPT; and, if a state ratifies or accedes to the NPT later than the expiry of the 180 days, then negotiations 'shall commence not later than the date of [their] deposit' of ratification or accession instrument. The point of commencement was on more permissive terms with regard to peaceful nuclear explosions (Art. V)—that is, 'as soon as possible after the [NPT] enters into force.'

[60] President Trump observed in reply to a reporter's question regarding possible future trilateral negotiation, 'We thought that we would do it first. I don't know if it's going to work out. But we would do it first and then we go to China together . . . Which, I think, works out probably better.' (Remarks by President Trump, July 29, 2020, *op. cit.* at 2 n. 2.) *Cf.* Trump (1987) at 233: 'There are times when you have to be aggressive, but there are also times when your best strategy is to lie back.'

[61] Ford, *op cit.*, therefore, was right when he observed that 'the language about negotiations needing to be "pursue[d] . . . in good faith" clearly leaves open the possibility that such negotiations might not take place, let alone succeed': Ford (2007) at 403. That possibility—that is, the possibility that negotiations might fail or never begin—is implicit in treaties that follow the more streamlined form and require, simply, that parties negotiate. Courts and tribunals that have been called upon to apply those treaties agree: see judgments and awards cited earlier, especially *Somalia v. Kenya*, Joint Declaration, Judges Gaja and Crawford, 2017 ICJ at 64, ¶4.

[62] See, for example, Meyer (2020) at 216.

[63] The drafters of the NPT in the 1960s were deliberate in leaving the complexity of arms control and disarmament to future negotiations.

Carlson (2019) at 101. See also the observation by US NPT negotiator Gerard Smith that it is 'obviously impossible to predict the exact nature and results of such negotiations.' *Military Implications of the Treaty on the Non-Proliferation of Nuclear Weapons*, Hearing Before the S. Comm. on Armed Services, 91st Cong. 121 (1969), (answer to question submitted by Sen. Thurmond).

[64] For a suggestion of the complexity, see case studies of six arms control negotiations, Carter (1989), *passim*.

[65] This conclusion follows from a general principle of legal construction: an allocation or reservation of discretion is not to deprive a binding term of effect. See *Certain Norwegian Loans (France v. Norway)*, 1957 ICJ at 9, 44, 66 (July 6) (separate opinion by Lauterpacht, J.).

four China and the NPT

[1] *National Security Strategy* (2022) at 20, 22.

[2] *National Security Strategy* (2017) at 25.

[3] *Id.* at 27.

[4] See generally Colby (2021).

[5] For a Chinese scholar's account, see Zhao (2019) at 371–94.

[6] Writers differ as to the precise goals motivating China's buildup but broadly agree that it is transformative. Particularly concerning is the possibility that China aims to achieve 'theater deterrence'—whereby its nuclear arsenal dissuades the US and US allies from defending against Chinese conventional attack. See Logan and Saunders (2023) at 15–16. Even analysts who have posited that China's buildup is for more limited aims concede that it is possible that 'China's pursuit of a stronger nuclear shield' might serve 'to enable conventional military operations or the non-nuclear escalation of a conflict': Cunningham (2023).

[7] 2022 Nuclear Posture Review at 4.

[8] *Id.*

[9] US Department of Defense, *Military and Security Developments Involving the People's Republic of China 2022* at 98.

[10] 2022 Nuclear Posture Review at 4.

[11] Remarks by President Trump Before Marine One Departure, The White House (July 29, 2020), Trump White House Archives.

[12] Interviewed by C.T. Lopez, 'Stratcom Commander Says U.S. Should Look to 1950s to Regain Competitive Edge,' *U.S. Department of Defense* (November 3, 2022): www.defense.gov/News/News-Stories/Article/Article/3209416/stratcom-commander-says-us-should-look-to-1950s-to-regain-competitive-edge/. See also Lucas Tomlinson's Tweet: https://twitter.com/LucasFoxNews/status/1588531359576121345

[13] Testimony of General Richard, March 8, 2022, reprinted (2022) 2(3) *Journal of Policy & Strategy* at 153. For original, see www.stratcom.mil/Portals/8/Documents/2022%20USSTRATCOM%20Posture%20Statement.pdf?ver=CUIoOCLyos9xe9C9I0XjMQ%3D%3D#page=3

[14] For my earlier reflections on the risk to strategic stability that China's nuclear weapons buildup presents, see quotations in Gertz (2022).

[15] Treaty on the elimination of their intermediate-range and shorter-range missiles, concluded December 8, 1987, entered into force June 1, 1988: 1657 UNTS at 2.

[16] See Trump White House Archives, *President Donald J. Trump to Withdraw the United States from the Intermediate-Range Nuclear Forces (INF) Treaty*, February 1, 2019: https://trumpwhitehouse.archives.gov/briefings-statements/president-donald-j-trump-withdraw-united-states-intermediate-range-nuclear-forces-inf-treaty/

[17] Tully (2024).

[18] Payne and Costlow (2022) at 4.

[19] Bender and O'Brien (2022).

[20] HMG (2021). As to which, see Grant (2021) at 49–55; Billingslea and Tully (2021).

[21] Schelling (1960) at 13. See also Congressional Commission on the Strategic Posture of the United States (2023) at 20–4.

[22] Grant (2021–22) at 63–4.

[23] See Ford (2007) at 408–9.

[24] See Bourantonis (1997) at 352.

[25] Grant (2021–22) at 15–16 with citations.

[26] *Id.* at 49–50.

[27] *Id.* at 50–3.

[28] Embassy of the People's Republic of China (2014) at ¶46.

[29] *Id.* at ¶47.

[30] *Id.* (emphasis added).

[31] Examples of states objecting on grounds of *fait accompli* are noted (without disapproval) here: *Territorial and Maritime Dispute (Nicar. v. Colom.)*, Judgment, 2012 ICJ at 624, 655, ¶79 (19 Nov.); CERD case, 2011 ICJ at 70, 79, ¶16; *Land and Maritime Boundary Between Cameroon and Nigeria*, 2002 ICJ at 439, ¶283. In the *Pulp Mills* case, Argentina objected that Uruguay's plans to carry out works on the River Uruguay would have constituted a *fait accompli*, a contention that the ICJ rejected—but this was after a negotiation period set in the relevant treaty had expired, a circumstance to which the Court drew attention: *Pulp Mills on the River Uruguay (Arg. v. Uru.)*, Judgment, 2010 ICJ at 14, 69–70, ¶153, 156–7 (April 20).

[32] See *South China Sea Arbitration (Philippines v. China)*, Award (2016) at ¶1177.

[33] As to bad faith delay and negotiation in international law, see *Lake Lanoux* Award, 12 UNRIAA 281, at 306–7 (November 16) (quoting the *Tacna-Arica* Award, 2 UNRIAA at 921 *et seq.* and *Railway Traffic between Lithuania and Poland*, Advisory Opinion, 1931 PCIJ Rep. (Ser. A/B) No. 42, at 108 *et seq.* (October 15). Bad faith delay has been considered in municipal law as well. For example, a public agency was found in breach of a duty to negotiate with a prospective contractor, where it delayed obligatory negotiations in order to '[explore] the possibility of soliciting more bids,' which is to say, it delayed in the hopes of the negotiating environment changing to its advantage over that lapse of time. *Banneker Ventures, LLC v. Graham*, 798 F.3d at 1119, 1131 (D.C. Cir. 2015).

[34] Courts applying municipal law have reached conclusions much to this effect. See, for example, 'a party might breach its obligation to bargain in good faith by unreasonably insisting on a condition outside the scope of the parties' preliminary agreement,' *A/S Apothekernes Lab'y. v. I.M.C. Chem. Grp., Inc.*, 873 F.2d at 155, 158 (7th Cir. 1989). The 'interject[ion] of new terms and conditions that were not part of' the agreed terms was also an issue in *Banneker Ventures. See* F.3d at 1131.

[35] *See* Ford (2020) at 3–4.

[36] *Electricity Company of Sofia and Bulgaria (Belgium v. Bulgaria)*, Interim Measures of Protection, Order, 1939 PCIJ (Ser. A/B) No. 79, at 199 (December 5); *South China Sea Arb.*, PCA Case No. 2013–19 at ¶¶1169–70. From another area of international practice, see Campolieti (2015) at 217–30.

[37] *South China Sea Arbitration*, Award (2016) at ¶1171.

[38] *Id.* at ¶1177 (emphasis added).

[39] *Id.* at ¶1178 (emphasis added).

[40] *Id.* at ¶1179 (emphasis added).

[41] See *South China Sea Arbitration*, Jurisdiction and Admissibility (2015) ¶413. Non-appearance by the respondent did not bar the proceedings (in accord with UNCLOS Annex VII, Art. 9). Award on Jurisdiction and Admissibility at ¶114.

[42] See *Gabčíkovo-Nagymaros Project*, 1997 ICJ at 66, ¶107.

[43] International law has no rules of general application that specify the manner or scope of preservation and production of evidence in negotiations or in judicial or arbitral procedures. Nevertheless, dispute settlement practice reflects the importance of maintaining a reliable evidentiary record, and particular procedural rules address the matter. For examples, see Grant (2021–22) at 52 n. 195. As to the necessity of sharing information in good faith, see Grant, Kent, and Trinidad (2022) at 395–6, 412–17, 421–5.

[44] If the view is accepted that negotiations require transparency and the exchange of information, then it is also relevant here that the Tribunal in

the *South China Sea Arbitration* found China to have failed to produce and transmit environmental impact assessments: *South China Sea Arbitration* (2016) at ¶991. Such conduct, if it reflects China's general practice, has particularly troubling implications for arms control. This is a further light in which non-compliance of Russia with the Open Skies Treaty, and the exiting of Russia and China from the International Partnership for Nuclear Disarmament Verification (IPNDV), are concerning. See Assistant Secretary of State (ISN) Ford, Testimony (2019) at 3. As to Russia's non-compliance, see US Department of Defense (2020).

45 The main international courts and tribunals, indeed, have the power to indicate measures of protection to prohibit such conduct in certain circumstances. See, for example, *Immunities and Criminal Proceedings (Eq. Guinea v. Fr.)*, Order, Request for the Indication of Provisional Measures, 2016 ICJ at 1169, ¶90 (December 7); *Seizure and Detention of Certain Documents and Data (Timor-Leste v. Austl)*, Order, Provisional Measures, 2014 ICJ at 157–9, ¶¶42–8 (March 3); *Delimitation of the Maritime Boundary (Ghana/Côte d'Ivoire)*, Provisional Measures, 2015 I.T.L.O.S. Case No. 23, at 163, ¶89 (April 25). See generally Miles (2017).

46 See Colby (2021), *passim*.

47 Sokolski and Beck (2022) at 5.

48 Gavin in Leffler and Westad (eds.) (2010) at 407. For the text of NASM 335 see www.discoverlbj.org/item/nsf-nsam335

49 China, Implementation of the Treaty on the Non-Proliferation of Nuclear Weapons (2019) at 21–2, ¶8.

50 Cong, Director-General of Department of Arms Control, Ministry of Foreign Affairs of the People's Republic of China (2020).

51 China, Implementation of the Treaty on the Non-Proliferation of Nuclear Weapons (2019) at ¶3(II).

52 Especially puzzling is China's interpolation into Art. VI of a qualification that it is acceptable for an NPT party to refrain from negotiating until the party judges that it is 'fair and equitable' to negotiate. Article VI reserves a great deal of discretion to states—for example, in respect of identifying 'effective measures.' It does not contain an equity qualification in regard to the size of negotiating states' nuclear arsenals, much less a self-judging equity qualification.

53 China, Implementation of the Treaty on the Non-Proliferation of Nuclear Weapons (2019) at ¶4. See also *id.* at ¶8.

54 It is not a contradiction to say that China has embarked upon an arms race, even though the US nuclear arsenal has remained almost static in quantity and has improved only marginally in quality: the arms race that most concerned the drafters of the NPT was not an 'action-reaction' loop either. See Trachtenberg, Dodge, and Payne (2021) at 11–21.

55 China, Implementation of the Treaty on the Non-Proliferation of Nuclear Weapons (2019) at ¶14.

56 *Id.* at ¶9.

57 *Id.* at ¶14.

58 *Id.* at ¶38.

59 *Id.*

60 European Parliament resolution of 15 December 2021 on the challenges and prospects for multilateral weapons of mass destruction arms control and disarmament regimes (2020/2001 (INI)) (2022/C 251/06).

61 *Id.* at ¶P.

62 *Id.* at ¶¶24, 25.

63 Sokolski and Beck (2022) at 8.

five What's Left to Negotiate?

1 Grant (2021–22) at 17–24; Grant (2021) at 17–19.

2 See, for example, NPT preamble: 'Declaring their intention to achieve at the earliest *possible* date the cessation of the nuclear arms race and to undertake effective measures in the *direction* of nuclear disarmament …' (emphasis added).

3 Grant (2021–22) at 24–7; Grant (2021) at 39. As for the phrase 'at an early date,' this does not prescribe a timeline. See, for example, Crawford (2017) at 459.

4 See 2022 Nuclear Posture Review at 17.

5 Sokolski and Beck (2022) at 5–6.

6 *Id.* at 5.

7 For example, missile defense is a topic that the US well might consider *not* negotiating. See Costlow, *Vulnerability Is No Virtue* (2022). See also Peters (2024). As Schelling observed, 'if we have to plan on the … assumption that the other side will strike first, 200 bombers safe against attack may be worth as much as 2000 that have only a 10 per cent chance of survival.' Schelling (1960) at 234.

8 Ford in Albertson (ed.) (2024) at 28.

9 See, for example, the 1998 UK Strategic Defence Review (SDR), ¶61, stating that '[t]ransparency about nuclear weapons holdings … plays a part in arms control,' quoted in *R. (on the application of Marchiori) v. Environment Agency* [2001] EWHC Admin 267 (March 19, 2001; Mr. Justice Turner); the US Senate Resolution of Ratification of the Start II Treaty (January 26, 1996), part c(1) on cooperative threat reductions, reprinted at Leich (ed.) (1996) at 275.

10 2022 Nuclear Posture Review at 13, 16, 18.

11 *Id.* at 17.

12 See, for example, Working paper submitted by New Zealand on behalf of the New Agenda Coalition (Brazil, Egypt, Ireland, Mexico, New Zealand, and South Africa), March 15, 2018, Prep. Com. for 2020 Rev. Con.: NPT/CONF/2020/PC.11/WP.13, ¶9, referring to Action 5 of 22 actions enumerated in the 2010 Rev. Con. action plan: 'further enhancing transparency and increasing mutual confidence.'

13 Sokolski (2015) at 5 n. 8.

14 *Id.*

15 For an exception, see Moon (2020) at 92–114, in which the author proposes confidence-building measures as part of a possible future renewal of cooperative efforts with Russia. Moon, now a consultant to the Stimson Center, served, *inter alia*, as EUCOM Theater Support Team Lead for the Defense Threat Reduction Agency (DTRA) (US).

16 China's aggressive acts against India suggest that China harbors territorial aims in that direction as well. See Joshi (2022).

17 Regarding the stabilizing effect of the post-1945 territorial settlement, which long was a reality in geopolitics and in international law, see Grant (2015) at 101–67. The view that the prohibition against territorial change through threat or use of force is the principal post-1945 legal rule has gained currency since Russia escalated its aggression against Ukraine in February 2022. See, for example, Brunk and Hakimi (2022) at 687–97.

18 Schelling and Halperin (1961) at 77–90.

19 The suggestion that confidence-building might help restore strategic stability relates to the analogous insight that a rush to nuclear arms *reductions* might well introduce risk if not conducted in a 'transparent, coordinated fashion': Sokolski (2015) at 5. Rapid change—even in a seemingly benign direction—upsets settled expectations and threat-perceptions.

20 See https://x.com/StateADS/status/1800590056744230983. See also British ministerial statements on the centrality of the 'international security environment' to nuclear arms control, for example, Minister of State, Foreign, Commonwealth and Development Office, Lord Ahmad of Wimbledon, HoL Hansard, vol. 809 col. 1268 (January 21, 2021); Graham Stuart, Minister of State, Foreign, Commonwealth and Development Office, HoC Hansard, vol. 718, cols. 149WH-150-WH (July 13, 2022).

21 Miller (2021) at 182. The US and USSR established the hotline in accordance with an MOU—the Memorandum of Understanding Regarding the Establishment of a Direct Communications Link (1963). The parties have updated their hotline arrangement in subsequent agreements. See Agreement of September 30, 1971; Agreement of July 17, 1984.

[22] The UK and USSR concluded a hotline agreement in 1967. Agreement concerning the establishment of a direct communication link between the Residence of the Prime Minister of the United Kingdom in London and the Kremlin (1967); Agreement on the Improvement of the Direct Communications Link between the Residence of the Prime Minister of the United Kingdom in London and the Kremlin (1987).

France and the USSR under a Joint Declaration of June 30, 1966 agreed to establish 'a direct line of communication' in regard, *inter alia*, to nuclear weapons issues: noted at Charpentier (ed.) (1966) at 919.

[23] See Joeck (2008) at 22.

[24] Sokolski and Beck (2022) at 13.

[25] See Joint Press Conference, Secretary of State Madeleine K. Albright and Foreign Minister Tang Jiaxuan, April 29, 1998: https://1997-2001.state.gov/statements/1998/980429.html. As to the long process that led to agreement to set up the US–China hotline, see Miller (2021) at 183 n. 21. The process started in 1971 with the Nixon–Kissinger diplomatic opening with China. *Id.*

[26] See Agreement on the Establishment of a Secure Defense Telephone Link between the Department of Defense, the United States of America and the Ministry of National Defense, the People's Republic of China (2008).

[27] Detsch (2021).

[28] Kine (2021).

[29] See Khan (2010) at 14. *Cf.* regarding underutilization of the India–Pakistan hotlines, Joeck (2008) at 35.

[30] See Miller (2021) at 181.

[31] Wang, Graff, and Dale-Huang (2024) at 19–21.

[32] *Id.* at 27.

[33] Morris and Marcrum (2022).

[34] *Id.*

[35] See 'Concerns Grow Over China Nuclear Reactors Shrouded in Mystery,' *Al Jazeera* (May 19, 2021): www.aljazeera.com/economy/2021/5/19/concerns-grow-over-china-nuclear-reactors-shrouded-in-mystery. For the earlier commitment, modest though it was, see IAEA INFCIRC/549 (March 16, 1998), at 3 (China's information note on 'the management of plutonium', dated December 1, 1997), www.iaea.org/sites/default/files/infcirc549.pdf

[36] Sokolski (2021) at 6–22.

[37] Kobayashi (2022).

[38] *Id.*

[39] GAR 48/75L, December 16, 1993, recommended the negotiation of a treaty to ban the production of fissile material for nuclear weapons. The

effort to negotiate such a treaty was endorsed by the NPT Rev. Cons in 1995, 2000, 2005, 2010, and in GAR 71/259, December 23, 2016.

40 See NPT/CONF.2020/PC.III/WP.2 (March 15, 2017), Working paper submitted by Australia, Austria, Canada, Denmark, Finland, Hungary, Ireland, the Netherlands, New Zealand, Norway and Sweden (the Vienna Group of Ten), ¶7.

41 China, Implementation of the Treaty on the Non-Proliferation of Nuclear Weapons (2019) at ¶23 (referring to Shannon report (CD/1299)).

42 Jenkins (2022).

43 For general background on the CD, see https://disarmament.unoda.org/conference-on-disarmament/

44 See Center for Arms Control and Non-Proliferation (2023).

45 S. 3010 – U.S.–China Nuclear Cooperation and Nonproliferation Act of 2016, introduced May 26, 2016, section 2, subparagraphs (9) and (11). See also, regarding Japan's Rokkasho plutonium recycling facility and its possible effects on fuel-enrichment activities in China and South Korea, Sokolski (2015) at 78–85. See also Gilinsky and Sokolski (2018).

46 2022 Nuclear Posture Review at 17.

47 LibreTexts Chemistry, 10.3: Isotopes of Hydrogen, https://chem.libretexts.org/Bookshelves/Inorganic_Chemistry/Map%3A_Inorganic_Chemistry_(Housecroft)/10%3A_Hydrogen/10.03%3A_Isotopes_of_Hydrogen

48 Nuclear Regulatory Commission (US) (2022).

49 See, for example, National Nuclear Safety Administration (US), Amended Record of Decision (2023).

50 Bergeron (2002) at 88.

51 For an accessible overview of the role of tritium in nuclear weapons, see Bergeron (2002) at 80–7.

52 *Id.*

53 *Id.* at 96.

54 See, for example, the Commerce Control List, Bureau of Industry and Security, Department of Commerce, 88 FR 56462 (August 18, 2023).

55 Prep. Com. for the 2020 Rev. Con., March 15, 2017, NPT/CONF.2020/PC.III/WP.2, Addressing 'Vienna issues,' Working paper submitted by Australia, Austria, Canada, Denmark, Finland, Hungary, Ireland, the Netherlands, New Zealand, Norway, and Sweden (the Vienna Group of Ten). Background Note 2, ¶1.

56 A number of countries, including Iraq and North Korea, have pursued covert nuclear weapons programs, notwithstanding IAEA safeguards. See Stricker in Goldberg (ed.) (2021) at 26.

57 Sokolski (2015) at 120.

58 See UNIDIR (2010) at 33.

[59] China, Implementation of the Treaty on the Non-Proliferation of Nuclear Weapons (2019) at ¶63.

[60] *Id.* at ¶¶43, 87.

[61] *Id.* at ¶87.

[62] Sokolski and Beck (2022) at 7.

[63] *Id.*

[64] S. 3010 – U.S.–China Nuclear Cooperation and Nonproliferation Act of 2016, introduced May 26, 2016, section 9(a) stating that 'it is the sense of Congress that the United States should encourage countries in East Asia to forego the commencement of new spent fuel reprocessing activities as part of a mutual effort to prevent the increased or expanded stockpiling of separated plutonium in the region.' Section 9(a).

[65] A proposal aired in Sokolski and Beck (2022) at 12.

[66] See, for example, US–Latvia Joint Statement on the New Clean Energy and Nuclear Security Collaboration (April 4, 2022): www.state.gov/joint-statement-on-the-new-clean-energy-and-nuclear-security-collaboration-under-the-foundational-infrastructure-for-responsible-use-of-small-modular-reactor-technology-first-initiative/

[67] See Fact Sheet: www.state.gov/nuclear-cooperation-memoranda-of-understanding-ncmou/

[68] Congressional Research Service (2016) at 1. *Cf. id.* at Summary. For an overview of the history of nuclear test bans, see Bellamy (2005) at 160–72.

[69] See www.un-ilibrary.org/content/books/9789210579780s005-c002

[70] *Id.*

[71] Treaty banning nuclear weapon tests in the atmosphere, in outer space and under water (1963).

[72] NPT Article X(2) provides that '[t]wenty-five years after the entry into force of the Treaty, a conference shall be convened to decide whether the Treaty shall continue in force indefinitely, or shall be extended for an additional fixed period or periods. This decision shall be taken by a majority of the Parties to the Treaty.'

[73] Congressional Research Service (2016) at 22.

[74] NPT/CONF.1995/32 (Part I) at 10.

[75] *Id.* at 9.

[76] See generally Ramaker, Mackby, Marshall, and Geil (2003).

[77] Conference on Disarmament doc. CD/1427 (August 22, 1996): https://documents-dds-ny.un.org/doc/UNDOC/GEN/G96/636/59/img/G9663659.pdf?OpenElement

[78] A/50/1027 (August 26, 1996): https://documents-dds-ny.un.org/doc/UNDOC/GEN/N96/219/23/img/N9621923.pdf?OpenElement

[79] GAR 50/245, September 17, 1996, ¶¶1 and 3.

80 See CD/1429.

81 NPT/CONF.2020/PC.III/WP.2 (March 15, 2017), Addressing 'Vienna issues,' Working paper submitted by Australia, Austria, Canada, Denmark, Finland, Hungary, Ireland, the Netherlands, New Zealand, Norway, and Sweden (the Vienna Group of Ten). See also Non-Proliferation and Disarmament Initiative States Working paper, reaffirming on behalf of 'a diverse cross-regional grouping of non-nuclear weapon States' their 'strong commitment to strengthening the nuclear test ban regime' and their 'strong concern that the [CTBT] still has not entered into force after 20 years': NPT/CONF.2020/PC.I/WP.3 (March 17, 2017).

82 For example, Tanzania, a long-standing stalwart of the Non-Aligned Movement, though abstaining from the General Assembly vote on adoption of the CTBT because (according to Tanzania) '[t]he treaty perpetuates the status quo by allowing the most technologically sophisticated nuclear-weapon States to continue with the vertical proliferation of nuclear arsenals through computer simulation' and because it 'provides for [no] further negotiations,' nevertheless voiced its endorsement of a test-ban as part of incremental progress toward eventual disarmament:

'Tanzania has all along been a staunch supporter and advocate of a comprehensive test-ban treaty (CTBT). We believed in the CTBT, and indeed we have always considered that it was *the only viable first step* leading to the total elimination of nuclear weapons' (emphasis added). Mr. Mwakawago (United Republic of Tanzania): A/50/PV.125 (September 10, 1996) at 1.

Pakistan, to give another example, 'consistently supported the objective of a comprehensive nuclear-test ban *as an essential step towards* nuclear disarmament and as a means of promoting nuclear non-proliferation.' Mr. Akram (Pakistan), *id.* at 7.

83 See, for example, Final Document of the 18th Non-Aligned Movement Summit of Heads of State and Government (2019) at 81 ¶244. For an earlier Non-Aligned Movement affirmation, see CTBT Organization Preparatory Commission, 'Non-aligned Movement Conference Stresses Importance of Comprehensive Nuclear-Test-Ban Treaty,' PI/2003/02 (2003).

84 Mr. Sha Zukang (China), A/50/PV.125 (September 10, 1996) at 10.

85 *Id.*

86 For a list of states having signed, including those having ratified, the CTBT, see https://treaties.un.org/Pages/showDetails.aspx?objid=0800000280049f7f&clang=_en

87 For detailed and critical discussion by a CTBT advocate, see Sykes (2017) at 195–210.

[88] As to which see Richard (2024).

[89] Under CTBT Article XIV, ratification by the states listed in Annex 2 to the Treaty (A/50/1027 at 48) is required in order for the CTBT to enter into force. Annex 2 lists 44 states, among them all five NPT nuclear-weapon states. Among the Annex 2 states, Egypt, Iran, and Israel (along with China and the US) have signed but not ratified the CTBT; North Korea, India, and Pakistan have not signed: www.nti.org/education-center/treaties-and-regimes/comprehensive-nuclear-test-ban-treaty-ctbt/#:~:text=After%20this%20ratification%2C%20nine%20Annex,Pakistan%2C%20and%20the%20United%20States

[90] Congressional Commission on the Strategic Posture of the United States (2023) at 17.

[91] *Id.* at 12.

[92] See US Department of State (2020) at 8.

[93] NPT/CONF.2020/PC.I/WP.3 (March 17, 2017), Comprehensive Nuclear-Test-Ban Treaty, Working paper submitted by the members of the Non-Proliferation and Disarmament Initiative (Australia, Canada, Chile, Germany, Japan, Mexico, the Netherlands, Nigeria, the Philippines, Poland, Turkey, and the United Arab Emirates).

[94] Report pursuant to 22 USC §2593a on Adherence to and Compliance with Arms Control, Nonproliferation, and Disarmament Agreements and Commitments (Department of State: June 2020).

[95] www.state.gov/adherence-to-and-compliance-with-arms-control-nonproliferation-and-disarmament-agreements-and-commitments/

[96] See Kenneth Adelman on definitional tussles in the 1980s with the Soviets: Adelman (1984) at 254.

[97] 2022 Nuclear Posture Review at 4 (emphasis added).

[98] www.foreign.senate.gov/imo/media/doc/05%2015%2019%20Intermediate-Range%20Nuclear%20Forces%20Treaty.pdf#page=47. Hearing before the Committee on Foreign Relations, United States Senate, 116th Cong. 1st Sess. (May 15, 2019) at 43. See also Congressional Commission on the Strategic Posture of the United States (2023) at 92–3.

[99] www.wsj.com/public/resources/documents/Russia.pdf?mod=article_inline#page=9

[100] www.wsj.com/public/resources/documents/Russia.pdf?mod=article_inline#page=9

[101] See, for example, China, Implementation of the Treaty on the Non-Proliferation of Nuclear Weapons (2019), ¶35. For the glossary, see *P5 Glossary of Key Nuclear Terms* (Beijing: China Atomic Energy Press, April 2015): https://2009-2017.state.gov/documents/organization/243293.pdf. As to China's involvement in the Glossary, see Berger (2014).

[102] NPT/CONF.2020/PC.III/8, ¶18.

[103] Bailey and Scheber (2011) at 15–16; Schneider (2019).

[104] www.state.gov/adherence-to-and-compliance-with-arms-control-nonproliferation-and-disarmament-agreements-and-commitments/

[105] The World Nuclear Association, as of October 2022, reports that China's civil nuclear generating capacity is 52,150 megawatts. https://world-nuclear.org/information-library/country-profiles/countries-a-f/china-nuclear-power.aspx. The Association reports that France's capacity is 61,370 megawatts: https://world-nuclear.org/information-library/country-profiles/countries-a-f/france.aspx

[106] Andrews-Speed (2020) at 23–46.

[107] Prep. Com. for the 2020 Rev. Con., March 15, 2017, NPT/CONF.2020/PC.I/WP.2 at 19, ¶4, Addressing 'Vienna issues,' Working paper submitted by Australia, Austria, Canada, Denmark, Finland, Hungary, Ireland, the Netherlands, New Zealand, Norway, and Sweden (the Vienna Group of Ten), Background note 5: nuclear safety. Including environmental safety and notification of coastal states of intended transit of nuclear materials.

[108] As to fuel diversion, see Ford and Countryman, Preface to Sokolski (ed.) (2021) at 1–2; Frank von Hippel in Sokolski (ed.) (2021) at 65–86. For a skeptical assessment, see Kristensen and Korda (2021) at 318–19.

[109] A connection might be drawn here to failures of transparency in other areas of Chinese practice, such as those that the Annex VII tribunal in the *South China Sea* arbitration determined to constitute breaches of China's obligations under UNCLOS, as to which see Chapter 4.

[110] See Congressional Research Service (2022).

[111] See, for example, 'DOE Announces Measures to Prevent China's Illegal Diversion of U.S. Civil Nuclear Technology for Military or Other Unauthorized Purposes,' US Department of Energy, October 11, 2018: www.energy.gov/articles/doe-announces-measures-prevent-chinas-illegal-diversion-us-civil-nuclear-technology

[112] Prep. Com. for the 2020 Rev. Con., NPT/CONF.2020/PC.III/WP.5 at 14 (March 15, 2019), Addressing 'Vienna issues,' Working paper submitted by Australia, Austria, Canada, Denmark, Finland, Hungary, Ireland, the Netherlands, New Zealand, Norway, and Sweden (the Vienna Group of Ten), Background note 3: export controls, ¶1. See also NPT/CONF.2020/WP.3/Rev.1 (June 20, 2022), at 16, background note 3.

[113] See Ford (2020) (delivered in Dr. Ford's capacity as Assistant Secretary for International Security and Nonproliferation and performing the duties of the Under Secretary of State for Arms Control and International Security). Concern over China's Military–Civil Fusion is not confined to US government circles. See, for example, Joshi (2022). The concern over Military–Civil Fusion extends past traditional defense considerations,

affecting as it does a range of US interests, such as patent protection: see Grant and Kieff (2021) at 262–70.

[114] Agreement for Cooperation concerning Peaceful Uses of Nuclear Energy (US) (2009).

[115] Protocol Additional to the Agreement between the United Arab Emirates and the International Atomic Energy Agency for the Application of Safeguards in Connection with the Treaty on the Non-Proliferation of Nuclear Weapons (2009). See www.iaea.org/sites/default/files/publications/documents/infcircs/2003/infcirc622a1.pdf

[116] See www.fdd.org/analysis/2023/09/21/letter-on-kingdom-of-saudi-arabias-request-for-uranium-enrichment/

[117] See Fortinsky (2023).

[118] As to the Chemical Weapons Convention, see www.opcw.org/about/mission; the Preparatory Commission to the Comprehensive Test Ban Treaty Organization, www.ctbto.org/our-mission; and the Biological Weapons Convention Implementation Support Unit, www.un.org/disarmament/biological-weapons/implementation-support-unit/relevant-activities-overseen-by-the-isu

[119] See Sokolski (2023).

[120] As to US withdrawal, see *DOD Statement on Open Skies Treaty Withdrawal* (May 21, 2020).

[121] See, for example, the Saudi proposal for a 'standing committee': *Negotiations of Articles I and II the NPT, Selected Documents*, vol. I (1961–66) at 532 (November 19, 1965).

[122] NPT/CONF.2020/PC.III/WP.36 (April 26, 2019), Security assurance, Working paper submitted by China, ¶3.

[123] See Tully (2024) regarding US intermediate-range missiles in the Pacific.

[124] See Sokolski (2022). For the 2011 ROK–Saudi agreement, which entered into force August 14, 2012, see UNTS reg. no. 54043.

[125] See Ryan (2022). Ryan, a retired major general in the Australian Defence Force, argues that basing US B-52s in Australia's Northern Territory allows Australia to allocate funds to conventional weapons which he identifies as 'more cost-effective ways to construct a strategic deterrent.'

[126] The argument the other way—that is, that nuclear weapons are the cheaper option—has a distinguished pedigree. It was supported by evidence in the 1950s, and the Eisenhower Administration, accordingly, in its 'New Look' defense strategy placed the emphasis on nuclear, rather than conventional, forces. See Dockrill (1996) at 48–71. The members of the Congressional Commission on the Strategic Posture of the United States concur that '[i]f the United States and its Allies and partners do not field sufficient conventional forces to [deter and defeat simultaneous Russian and Chinese aggression in Europe and Asia], U.S. strategy would

need to be altered to increase reliance on nuclear weapons.' Congressional Commission (2023) at viii.

[127] See Bowman and Thompson (2021).

[128] Gavin in Leffler and Westad (eds.) (2010) at 405–6.

[129] Gavin in Leffler and Westad (eds.) (2010) at 410, quoting 'Summary of NPT Issues Paper' (January 28, 1969), I-2, box H-019, NSC Meetings File, Nixon Presidential Materials Project.

[130] Gavin at 399.

Conclusion: An NPT Future and Bringing Realists Back to Arms Control

[1] HoL Hansard, vol. 809, col. 1270 (January 21, 2021).

[2] *Id.*

[3] Søe (1995) at 497–8.

[4] For a digital archive relating to the cruise missile ban movement, see www.reportdigital.co.uk/archive-photos/1980s/cnd-greenham-common-and-cruisewatch-1980s.html

[5] West German conservatives at the time expressed privately their concern that, if the anti-missile activists and their supporters in the SPD (Socialist Party) were to prevail, the Atlantic Alliance would collapse. See Herf (1991) at 218.

[6] See generally Cronin (2014). As to the role that President Reagan and his Secretary of State, George P. Shultz, played, see Kieninger (2020). See also McNeill (1985) at 62–3.

[7] Treaty between the United States of America and the Union of Soviet Socialist Republics on the elimination of their intermediate-range and shorter-range missiles (1987). See further Leich (1988) at 341–50.

[8] See Congressional Commission on the Strategic Posture of the United States (2023) at 78.

[9] Sokolski and Grant (2022).

[10] President Biden, on the occasion of the 10th NPT Rev. Con., called on China to observe its obligations under the NPT: President Biden Statement Ahead of the 10th Review Conference of the Treaty on the Non-Proliferation of Nuclear Weapons (August 1, 2022): www.whitehouse.gov/briefing-room/statements-releases/2022/08/01/president-biden-statement-ahead-of-the-10th-review-conference-of-the-treaty-on-the-non-proliferation-of-nuclear-weapons/

[11] Billingslea (2020).

[12] Pompeo and Billingslea (2021).

[13] See Cruz (2021); ICAN (2021).

[14] Kissinger (1984).

References

Primary materials

Acheson–Lilienthal Report, https://history.state.gov/milestones/1945-1952/baruch-plans

Bettauer, R., Deputy Legal Adviser of the US Department of State, Address Before Lawyer's Committee on Nuclear Policy (October 10, 2006), https://2009-2017.state.gov/s/l/2006/98879.htm

China, Implementation of the Treaty on the Non-Proliferation of Nuclear Weapons of the People's Republic of China, Report submitted by China, Preparatory Committee for the 2020 Review Conference, NPT/CONF.2020/PC.III/8 (April 29, 2019), https://documents.un.org/doc/undoc/gen/n19/124/15/pdf/n1912415.pdf?token=PUHgKqdUljA0WitiEv&fe=true

Cong, F., Director-General of Department of Arms Control, Ministry of Foreign Affairs of the People's Republic of China, interview in *Kommersant* (October 16, 2020), www.fmprc.gov.cn/mfa_eng/wjbxw/t1824545.shtml

Congressional Commission on the Strategic Posture of the United States, *America's Strategic Posture* (Final Report of the Commission) (October 2023)

Embassy of the People's Republic of China in the Republic of Fiji, Position Paper of the Government of the People's Republic of China on the Matters of Jurisdiction in the South China Sea Arbitration Initiated by the Republic of the Philippines, ¶46 (December 7, 2014), www.mfa.gov.cn/ce/cefj/eng/topic/nhwt/t1372318.htm

Ford, C.A., Assistant Secretary of State (ISN), Testimony before the US Senate Foreign Relations Committee (December 3, 2019), 3, www.foreign.senate.gov/imo/media/doc/120319_Ford_Testimony.pdf

Hansard, House of Commons (UK)

Hansard, House of Lords (UK)

HMG (Her Majesty's Government), *Global Britain in a competitive age. The Integrated Review of Security, Defence, Development and Foreign Policy* (March 2021)

IAEA Information Circulars, https://www.iaea.org/publications/documents/infcircs

Jenkins, Amb. B.D., Under Secretary for Arms Control and International Security, Remarks, Washington, DC (May 26, 2022), www.state.gov/priorities-regarding-the-new-and-emerging-challenges-to-international-security/

Labour's Manifesto (June 2024), https://labour.org.uk/change/strong-foundations/

Leich, M.N. (ed.), 'Contemporary Practice of the United States Relating to International Law: Arms Control and Disarmament' (1988) 82 *American Journal of International Law* 341 and (1996) 90:272

National Defense Strategy of the United States of America (2022) including the 2022 Nuclear Posture Review and the 2022 Missile Defense Review: https://media.defense.gov/2022/Oct/27/2003103845/-1/-1/1/2022-NATIONAL-DEFENSE-STRATEGY-NPR-MDR.PDF

National Security Strategy of the United States of America (December 2017), p 25: https://trumpwhitehouse.archives.gov/wp-content/uploads/2017/12/NSS-Final-12-18-2017-0905.pdf#page=35

Negotiations of Articles I and II of the NPT. Selected Documents, vol. 2 (1966–68) (NATO)

Nixon Presidential Materials Project (NSC Meetings File, 1969)

NPT Conference materials, https://disarmament.unoda.org/wmd/nuclear/npt-review-conferences/

Office of General Counsel, Department of Defense, *Law of War Manual* (Arlington, VA: December 2016)

Public Papers of the Presidents: John F. Kennedy, 1963

Sanders, J., NPT Article IV, Statement at the Third Session of the Preparatory Committee for the 2005 Review Conference of the NPT, New York (April 29, 2004), https://2001-2009.state.gov/t/isn/rls/rm/32292.htm

Sanders, J., US Special Rep. of the President for the Nonproliferation of Nuclear Weapons, Statement to the 2005 NPT Review Conference Of the Treaty on the Nonproliferation of Nuclear Weapons: U.S. Implementation of Article VI and the Future of Nuclear Disarmament (May 20, 2005), https://2001-2009.state.gov/t/isn/rls/rm/46603.htm

Trump White House Archives, https://trumpwhitehouse.archives.gov

United Kingdom 1998 Strategic Defence Review (SDR)

United Nations Disarmament Yearbook

US Department of Defense, *Military and Security Developments Involving the People's Republic of China 2022*, https://media.defense.gov/2022/Nov/29/2003122279/-1/-1/1/2022-MILITARY-AND-SECURITY-DEVELOPMENTS-INVOLVING-THE-PEOPLES-REPUBLIC-OF-CHINA.PDF#page=118

US Department of Defense Law of War Manual (US Department of Defense: updated July 2023), https://media.defense.gov/2023/Jul/31/2003271432/-1/-1/0/DOD-LAW-OF-WAR-MANUAL-JUNE-2015-UPDATED-JULY%202023.PDF

US Department of State, *Executive Summary of Findings on Adherence to and Compliance with Arms Control, Nonproliferation, and Disarmament Agreements and Commitments* (April 2020)

Wilson Center Digital Archive (NPT materials from French foreign ministry)

Treaties

Protocol for the Prohibition of the Use in War of Asphyxiating, Poisonous or Other Gases, and of Bacteriological Methods of Warfare, June 17, 1925, entered into force February 8, 1928: 94 LNTS 65

Statute of the International Atomic Energy Agency, concluded October 26, 1956, entered into force July 29, 1957: 276 UNTS 3

Memorandum of Understanding Regarding the Establishment of a Direct Communications Link, concluded June 20, 1963: 472 UNTS 163

Treaty banning nuclear weapon tests in the atmosphere, in outer space and under water ('Partial Test Ban Treaty'), concluded August 5, 1963, entered into force October 10, 1963: 480 UNTS 43

Joint Declaration of June 30, 1966 (France–USSR), noted at Charpentier, J. (ed.), 'Pratique française du Droit International' (1966) 12 *Annuaire français de droit international* 919

Agreement concerning the establishment of a direct communication link between the Residence of the Prime Minister of the United Kingdom in London and the Kremlin, concluded August 25, 1967: 632 UNTS 49

Treaty on the Non-Proliferation of Nuclear Weapons (NPT), opened for signature July 1, 1968; entered into force March 5, 1970: 21 UST 483, 729 UNTS 161

Memorandum of Understanding between the United States of America and the Union of Soviet Socialist Republics Regarding the Establishment of a Direct Communications Link, signed September 30, 1971: 806 UNTS 402

Convention on the prohibition of the development, production and stockpiling of bacteriological (biological) and toxin weapons and their destruction, concluded April 10, 1972, entered into force March 25, 1975: 1015 UNTS 163

United Nations Convention on the Law of the Sea (UNCLOS), opened for signature December 10, 1982, entered into force November 14, 1994: 1833 UNTS 397

Agreement of July 17, 1984 (relating to September 30, 1971 US–USSR Direct Communications Link Memorandum): 2193 UNTS 51

Agreement on the Improvement of the Direct Communications Link between the Residence of the Prime Minister of the United Kingdom in London and the Kremlin, concluded March 31, 1987: 1655 UNTS 393

Treaty between the United States of America and the Union of Soviet Socialist Republics on the elimination of their intermediate-range and shorter-range missiles (with memorandum of understanding, protocol on procedures governing the elimination of the missile systems, protocol regarding inspections and annex thereto, site diagrams and photographs, exchange of notes and agreed minute dated May 12, 1988 and exchange of notes dated May 28 and 29, 1988): signed December 8, 1987; entered into force June 1, 1988: 1657 UNTS 5

Convention on the Prohibition of the Development, Production, Stockpiling and Use of Chemical Weapons and on their Destruction, concluded January 13, 1993, entered into force April 19, 1997: 1974 UNTS 45, 1975 UNTS 3

Treaty on Open Skies, signed March 24, 1992, entered into force January 1, 2002: UK Treaty Series No. 27 (2002)

Convention on the Prohibition of the Use, Stockpiling, Production and Transfer of Anti-Personnel Mines and on their Destruction, concluded September 18, 1997, entered into force March 1, 1999: 2056 UNTA 211

Agreement on the Establishment of a Secure Defense Telephone Link between the Department of Defense, the United States of America and the Ministry of National Defense, the People's Republic of China, concluded February 29, 2008: www.state.gov/wp-content/uploads/2019/02/08-229-China-Telecommunication-Link.EnglishOCR.pdf

New Start Treaty (Treaty between the United States of America and the Russian Federation on Measures for the Further Reduction and Limitation of Strategic Offensive Arms), signed April 8, 2010, entered into force February 5, 2011: TIAS 11-205; extended January 2, 2021: TIAS 21-203

Convention on Cluster Munitions, concluded December 3, 2008, entered into force August 1, 2010: 2688 UNTS 92

Protocol Additional to the Agreement between the United Arab Emirates and the International Atomic Energy Agency for the Application of Safeguards in Connection with the Treaty on the Non-Proliferation of Nuclear Weapons, signed April 8, 2009; entered into force December 20, 2010: www.iaea.org/sites/default/files/publications/documents/infcircs/2003/infcirc622a1.pdf

Protocol Additional to the Agreement between the Government of India and the International Atomic Energy Agency for the Application of Safeguards to Civilian Nuclear Facilities, Vienna, signed May 15, 2009; entered into force July 25, 2014: 3077 UNTS 257

Agreement for Cooperation concerning Peaceful Uses of Nuclear Energy (US–UAE), concluded May 21, 2009: www.govinfo.gov/content/pkg/CDOC-111hdoc43/pdf/CDOC-111hdoc43.pdf

Agreement between the Government of the Republic of Korea and the Government of the Kingdom of Saudi Arabia for cooperation in the peaceful uses of nuclear energy, concluded November 15, 2011, entered into force August 14, 2012: UNTS reg. no. 54043

Final Agreement for the settlement of the difference described in the United Nations Security Council Resolutions 817 (1993) and 845 (1993), the termination of the Interim Accord of 1995, and the establishment of a strategic partnership between the Parties, signed at Prespa, June 17, 2018, and entered into force February 12, 2019: UNTS reg. no. 55707

Treaty on the Prohibition of Nuclear Weapons (TPNW), concluded July 7, 2017; entered into force January 22, 2021: UNTS reg. no. 56487

Cases

The Case of the 'S.S. Lotus' (France v. Turkey), 1927 PCIJ Ser. A., No. 10 (September 7)

Railway Traffic between Lithuania and Poland, Advisory Opinion, 1931 PCIJ Rep. Ser. A/B No. 42 (October 15)

Conditions of Admission of a State to Membership in the United Nations (Article 4 of the Charter), Advisory Opinion, 1948 ICJ 57 (May 28)

Lake Lanoux Award, XII UNRIAA 281 (November 16, 1957)

Temple of Preah Vihear (Cambodia v. Thailand), Judgment, 1962 ICJ 6, 57 (June 15) (separate opinion, Fitzmaurice, J.)

North Sea Continental Shelf (Germany/Denmark; Germany/Netherlands), Judgment, 1969 ICJ 3 (February 20)

Border and Transborder Armed Actions (Nicaragua v. Honduras), Jurisdiction and Admissibility, 1988 ICJ 69 (December 20)

Legality of the Threat or Use of Nuclear Weapons, Advisory Opinion, 1996 ICJ 66 (July 8)

Gabčíkovo-Nagymaros Project (Hungary v. Slovakia), Judgment, 1997 ICJ 7 (September 25)

Land and Maritime Boundary between Cameroon and Nigeria (Cameroon v. Nigeria: Equatorial Guinea intervening), Judgment, 2002 ICJ 303 (October 10)

Armed Activities on the Territory of the Congo (New Application: 2002) (Democratic Republic of Congo v. Rwanda), Judgment, 2006 ICJ 6

Maritime Delimitation in the Indian Ocean (Somalia v. Kenya), Preliminary Objections, Judgment, 2017 ICJ 3 (February 2)

Application of the Int'l Convention on the Elimination of All Forms of Racial Discrimination (Georgia v. Russian Federation), Preliminary Objections, Judgment, 2011 ICJ 70 (April 1)

Application of the Interim Accord of 13 September 1995 (the former Yugoslav Republic of Macedonia v. Greece), 2011 ICJ p 644 (December 5)

South China Sea Arbitration (Philippines v. China), PCA Case No. 2013-19, Award on Jurisdiction and Admissibility (October 29, 2015)

Obligations concerning Negotiations relating to Cessation of the Nuclear Arms Race and to Nuclear Disarmament (Marshall Islands v. India), 2016 ICJ 255 (October 5)

Obligations concerning Negotiations relating to Cessation of the Nuclear Arms Race and to Nuclear Disarmament (Marshall Islands v. Pakistan), 2016 ICJ 552 (October 5)

Obligations concerning Negotiations relating to Cessation of the Nuclear Arms Race and to Nuclear Disarmament (Marshall Islands v. United Kingdom), Judgment, 2016 ICJ 833 (October 5)

South China Sea Arbitration (Philippines v. China), PCA Case No. 2013-19, Award (July 12, 2016)

Application of the Int'l Convention for the Suppression of the Financing of Terrorism and of the Int'l Convention on the Elimination of All Forms of Racial Discrimination (Ukraine v. Russian Federation), Preliminary Objections, Judgment, 2019 ICJ 558 (November 8)

Appeal relating to the Jurisdiction of the ICAO Council (Bahrain, Egypt and United Arab Emirates v. Qatar), Judgment, 2020 ICJ 172 (July 14)

Resolutions

GAR 1(I) ('Establishment of a Commission to Deal with the Problems Raised by the Discovery of Atomic Energy'), January 24, 1946

GAR 2028 (XX) ('Non-proliferation of nuclear weapons'), November 19, 1965

GAR 2373 (XXII) ('Treaty on the Non-Proliferation of Nuclear Weapons'), June 12, 1968

GAR 47/52 A ('Preparatory Committee for the 1995 Conference of the States Parties to the Treaty on the Non-Proliferation of Nuclear Weapons'), December 9, 1992

Legality of the Use by a State of Nuclear Weapons in Armed Conflict, Request for Advisory Opinion, World Health Assembly resolution of May 14, 1993

GAR 48/75L ('Prohibition of the Production of Fissile Material for Nuclear Weapons or Other Nuclear Explosive Devices'), December 16, 1993

GAR 49/75 K ('Request for an advisory opinion from the International Court of Justice on the legality of the threat or use of nuclear weapons'), December 15, 1994

GAR 50/245 ('Comprehensive nuclear-test-ban treaty'), September 10, 1996

SCR 1887 (2009), September 24, 2009

GAR 71/258 (2016) ('Taking forward multilateral nuclear disarmament negotiations: Statement of financial implications'), December 23, 2016

GAR 71/259 ('Treaty banning the production of fissile material for nuclear weapons or other nuclear explosive devices'), December 23, 2016

GAR 75/40 ('Treaty on the Prohibition of Nuclear Weapons'), December 7, 2020

European Parliament resolution of 15 December 2021 on the challenges and prospects for multilateral weapons of mass destruction arms control and disarmament regimes (2020/2001 (INI)) (2022/C 251/06)

Other instruments

Floyd D. Spence National Defense Authorization Act for Fiscal Year 2001, sec. 1041 (Revised Nuclear Posture Review) (US)

Guiding Principles Applicable to Unilateral Declarations of States Capable of Creating Legal Obligations, ILC Rep. on the Work of its Fifty-Eighth Session, 2006 ILC Yearbook 161

P5 Joint Statement on the TPNW (October 24, 2018), www.gov.uk/government/news/p5-joint-statement-on-the-treaty-on-the-nonproliferation-of-nuclear-weapons

Final Document of the 18th Non-Aligned Movement Summit of Heads of State and Government, Baku (October 25–26, 2019), NAM 2019/CoB/Doc.1

US Department of Defense *Statement on Open Skies Treaty Withdrawal* (May 21, 2020), www.defense.gov/News/Releases/Release/Article/2195239/dod-statement-on-open-skies-treaty-withdrawal/ (effective November 22, 2020)

Russian Federation, Foreign Ministry statement on the withdrawal of the Russian Federation from the Treaty on Open Skies (December 18, 2021), www.mid.ru/en/foreign_policy/news/1790948/

National Nuclear Safety Administration (US), Amended Record of Decision for the Production of Tritium in Commercial Light Water Reactors (September 14, 2023): 88 FR 63099

Secondary works

Adelman, K.L., 'Arms Control With and Without Agreements', (1984) 63(2) *Foreign Affairs* 240

Akande, D., 'Nuclear Weapons, Unclear Law? Deciphering the *Nuclear Weapons* Advisory Opinion of the International Court', (1998) 68 *British Yearbook of International Law* 165

Akita, H., 'How China Continues to Lose Friends in Central and Eastern Europe', *Nikkei Asia* (November 3, 2022)

Andrews-Speed, P., 'The Governance of Nuclear Power in China', (2020) 13(1) *Journal of World Energy Law & Business* 23

Arimatsu, L., 'Transformative Disarmament: Crafting a Roadmap for Peace', (2021) 97 *International Law Study Series US Naval War College* 833

Arms Control Association, *Nuclear Weapons: Who Has What at a Glance* (January 2022), www.armscontrol.org/factsheets/Nuclearweaponswhohaswhat

Bailey, K. and Scheber, T., *The Comprehensive Test Ban Treaty: An Assessment of the Benefits, Costs, and Risks* (Fairfax, VA: National Institute for Public Policy, March 2011)

Bellamy, I., *Curbing the Spread of Nuclear Weapons* (Manchester: Manchester University Press, 2005)

Bender, B. and O'Brien, C., 'Top GOP Hawks Warn Biden against Nuclear Cuts', *Politico* (January 13, 2022)

Berger, A., *The P5 Nuclear Dialogue. Five Years On*, RUSI Occasional Paper (July 2014)

Bergeron, K.D., *Tritium on Ice. The Dangerous New Alliance of Nuclear Weapons and Nuclear Power* (Cambridge: MIT Press, 2002)

Bezerra, M., 'IAEA Additional Protocol and its Impact on Relations between Brazil and Argentina', (2021) 12 *Journal of the Russian Academy of Science* 39

Billingslea, M.S., 'Arms Control and the New Start Treaty' (December 8, 2020), https://nipp.org/information_series/billingslea-marshall-arms-control-and-the-new-start-treaty-information-series-no-472/

Billingslea, M.S. and Tully, R., 'The U.K.'s Response to Russian and Chinese Aggression', *RealClearDefense* (March 22, 2021)

Bolton, J.R., 'The NPT: A Crisis of Non-Compliance', Statement to the Third Session of the Preparatory Committee for the 2005 Review Conference of the NPT, New York (April 27, 2004)

Bourantonis, D., 'The Negotiation of the Non-Proliferation Treaty, 1965–1968', (1997) 19(2) *International History Review* 347

Bowman, B. and Thompson, J., 'Russia and China Seek to Tie America's Hands in Space. Biden Should Avoid the Treaty Trap Set by Moscow and Beijing', *Foreign Policy* (March 31, 2021)

Brunk, I. and Hakimi, M., 'Russia, Ukraine, and the Future World Order', (2022) 116(4) *AJIL* 687

Campolieti, F., 'The Rule of Non-Aggravation of the Dispute in ICSID Arbitration Practice', (2015) 30(1) *ICSID Review—Foreign Investment Law Journal* 217

Carlson, J., 'Is the NPT Still Relevant? How to Progress the NPT's Disarmament Provisions', (2019) 2(1) *Journal for Peace and Nuclear Disarmament* 97

Carter, A., *Success and Failure in Arms Control Negotiations* (Oxford: Oxford University Press, 1989)

Center for Arms Control and Non-Proliferation, *Fissile Material Cutoff Treaty (FMCT)*, https://armscontrolcenter.org/wp-content/uploads/2023/05/Fissile-Material-Cutoff-Treaty-FMCT-Fact-Sheet.pdf (2023)

Colby, E.A., *The Strategy of Denial. American Defense in an Age of Great Power Conflict* (New Haven, CT: Yale University Press, 2021)

Congressional Research Service, *Comprehensive Nuclear-Test-Ban Treaty: Background and Current Developments* (September 1, 2016)

Congressional Research Service, *U.S. Export Controls and China* (updated March 24, 2022), https://crsreports.congress.gov/product/pdf/IF/IF11627

Cormier, M. and Hood, A., 'Australia's Reliance on US Extended Nuclear Deterrence and International Law', (2017) 13 *Journal of International Law & International Relations* 3

Costlow, M.R., 'Review: David A. Cooper, *Arms Control for the Third Nuclear Age: Between Disarmament and Armageddon* (Washington, DC: Georgetown University Press, 2021)', (2022) 2(3) *Journal of Policy & Strategy* 148

Costlow, M.R., *Vulnerability Is No Virtue and Defense Is No Vice: The Strategic Benefits of Expanded U.S. Homeland Missile Defense* (National Institute for Public Policy, Occasional Paper Vol. 2, No. 9, September 2022)

Crawford, J.R., 'Chance, Order, Change: The Course of International Law', (2013) 365 *Hague Academy Collected Courses* 113

Crawford, J.R., 'International Law and the Problem of Change: A Tale of Two Conventions', (2017) 49 *Victoria University of Wellington Law Review* 459

Cronin, J.E., *Global Rules: America, Britain and a Disordered World* (New Haven, CT: Yale University Press, 2014)

Cruz, T., 'We Should Enact Sanctions on China's Defense Industries Unless They Comply With Their Treaty Obligations to Halt Their Buildup', *Twitter* (May 18, 2021), https://twitter.com/sentedcruz/status/1394752627368513538

Cunningham, F.S., 'The Unknowns About China's Nuclear Modernization Program', *Arms Control Today* (June 2023)

Detsch, J., 'Biden Looks for Defense Hotline With China', *Foreign Policy* (May 10, 2021)

Dockrill, S., *Eisenhower's New-look National Security Policy, 1953–61* (Basingstoke, UK: Palgrave Macmillan, 1996)

Dunworth, T., 'Pursuing "Effective Measures" Relating to Nuclear Disarmament: Ways of Making a Legal Obligation a Reality', (2016) 97 *International Review of the Red Cross* 601

ElBaradei, M., *Addressing Verification Challenges* (October 16, 2006), www.iaea.org/newscenter/statements/addressing-verification-challenges

Erästö, T., 'The NPT and the TPNW: Compatible or Conflicting Nuclear Weapons Treaties?', *SIPRI WritePeace blog* (March 6, 2019), www.sipri.org/commentary/blog/2019/npt-and-tpnw-compatible-or-conflicting-nuclear-weapons-treaties

Evangelista, M., 'Transnational Organization and the Cold War', in M.P. Leffler and O.A. Westad (eds.) *The Cambridge History of the Cold War*, vol. III (Cambridge: Cambridge University Press, 2010) 400

Ford, C.A., 'Debating Disarmament: Interpreting Article VI of the Treaty on the Non-Proliferation of Nuclear Weapons', (2007) 14(3) *Nonproliferation Review* 401

Ford, C.A., *Nuclear Technology Rights and Wrongs: The Nuclear Nonproliferation Treaty, Article IV, and Nonproliferation,* Research Report (May 1, 2010) at p 237, www.npec-web.org/Essays/20090601-Ford-NuclearRightsAndWrongs.pdf

Ford, C.A., 'The NPT Regime and the Challenge of Shaping Proliferation Behavior', in P.R. Lavoy and J.J. Wirtz (eds.) *Over the Horizon Proliferation Threats* (Redwood City, CA: Stanford University Press, 2012) 179

Ford, C.A., *The PRC's 'Military-Civil Fusion' Strategy Is a Global Security Threat*, US Department of State Dipnote (March 16, 2020)

Ford, C.A., 'Competitive Strategy vis-à-vis China and Russia: A View from the "T Suite"', (2020) 1(6) *ACIS Papers* 2, at 3–4

Ford, C.A., *Assessing the Biden Administration's 'Big Four' National Security Guidance Documents*, NIPP Occasional Paper, Vol. 3, No. 1 (January 2023) (Fairfax, VA: National Institute Press, 2023)

Ford, C.A., 'Dead or Deferred? Nuclear Arms Control in an Age of Revisionism', in M. Albertson (ed.) *Aligning Arms Control with the New Security Environment* (Center for Global Security Research, Lawrence Livermore National Laboratory, May 2024) 25

Ford, C.A. and Countryman, T., Preface to Sokolski, H. (ed.), *China's Civil Nuclear Sector: Plowshares to Swords?* (NPEC Occasional Paper 2102, March 2021)

Fortinsky, S., 'Saudi Crown Prince on Iran Acquiring Nuclear Weapons: "If they get one, we have to get one"', *The Hill* (September 20, 2023)

Francis, S., 'Starmer Says He Is Prepared to Use Nuclear Weapons', *BBC News* (June 3, 2024), www.bbc.co.uk/news/articles/czvvy0ppdxko

Freeman, A.V., 'The Development of International Co-Operation in the Peaceful Use of Atomic Energy', (1960) 54(2) *American Journal of International Law* 383

Gavin, F.J., 'Nuclear Proliferation and Non-Proliferation During the Cold War', in M.P. Leffler and O.A. Westad (eds.) *The Cambridge History of the Cold War*, vol. II (Cambridge: Cambridge University Press, 2010) 395

Gertz, B., 'Top Chinese Military Figure Calls Sharp Buildup "Appropriate"', *Washington Times* (June 14, 2022), www.washingtontimes.com/news/2022/jun/14/gen-wei-fenghe-top-chinese-military-figure-calls-s/

Gilinsky, V. and Sokolski, H., 'Make US–Japanese Nuclear Cooperation Stable Again: End Reprocessing', *Bulletin of the Atomic Scientists* (June 27, 2018)

Graham, K., 'The TPNW Conference of Parties: What Is to Be Discussed?' (2020) 3 *Journal for Peace and Nuclear Disarmament* 234

Grant, T.D., 'How To Recognise a State (and Not): Some Practical Considerations', in C. Chinkin and F. Baetens (eds.) *Sovereignty, Statehood and State Responsibility: Essays in Honour of James Crawford* (2015) 192

Grant, T.D., *Aggression Against Ukraine: Territory, Responsibility, and International Law* (Macmillan, 2015)

Grant, T.D., 'China's Nuclear Build-Up and Article VI NPT: Legal Text and Strategic Challenge', National Institute for Public Policy Occasional Paper, vol. 1, no. 11 (November 2021)

Grant, T.D., 'Article VI of the Treaty on the Non-Proliferation of Nuclear Weapons (NPT): The Pursuit of Negotiations and China's Pursuit of Arms', (2021–22) 31 *Journal of Transnational Law & Policy* 1

Grant, T.D., 'The ILC's Draft Conclusions on Customary International Law and the Treaty on the Prohibition of Nuclear Weapons (TPNW): Lawmaking and its Limits in a Nuclear Age', (2022) 13(2) *George Mason International Law Journal* 1

Grant, T.D. and Kieff, F.S., 'Great Powers and New Risks: What Businesses and Regulators Should Know about China's Strategic Ambitions', (February 2021) 65(2) *Orbis* 257

Grant, T.D., Kent, A., and Trinidad, J., 'Better Late than Never? The Environmental Impact Assessment and its Timing and Function', (2022) 39(3) *Wisconsin Journal of International Law* 391

Groves, S. and Bromund, T., 'The Ottawa Mine Ban Convention: Unacceptable on Substance and Process', www.heritage.org/global-politics/report/the-ottawa-mine-ban-convention-unacceptable-substance-and-process (Washington, DC: Heritage Foundation, December 13, 2010)

Harries, M., *Disarmament as Politics: Lessons from the Negotiation of NPT Article VI* (London: Chatham House, May 2015)

Heft, J., *War by Other Means: Soviet Power, West German Resistance, and the Battle of the Euromissiles* (New York: The Free Press, 1991)

Hertogen, A., 'Letting *Lotus* Bloom', (2016) 26(4) *European Journal of International Law* 901

Highsmith, N. and Stewart, M., 'The Nuclear Ban Treaty: A Legal Analysis', (2018) 60(1) *Survival* 129

Hood, A. and Cormier, M., 'Can Australia Join the Nuclear Ban Treaty without Undermining ANZUS?', (2020) 44 *Melbourne University Law Review* 132

Hooper, R., 'The Changing Nature of Safeguards', *IAEA Bulletin*, no. 45/1 (June 2003), at 11, www.iaea.org/Publications/Magazines/Bulletin/Bull 451/article2.pdf

ICAN, 'Divestment and Nuclear Weapons' (April 16, 2020), www.icanw.org/divestment_and_nuclear_weapons

ICAN, 'China's Nuclear Buildup Violates International Law' (November 12, 2021), www.icanw.org/china_nuclear_buildup

International Institute for Strategic Studies (IISS), 'Poland's Bid to Participate in NATO Nuclear Sharing', 29(7) *Strategic Comments* (September 11, 2023)

Joeck, N., 'The Indo-Pakistani Nuclear Confrontation: Lessons from the Past, Contingencies for the Future', Nonproliferation Policy Education Center Working Paper Series (September 2008)

Jonas, D.S. and Braunstein, A.E., 'What's Intent Got to Do With It? Interpreting "Peaceful Purpose" in Article IV.1 of the NPT', (2018) 32 *Emory International Law Review* 351

Joshi, M., 'China's Military-Civil Fusion Strategy, the US Response, and Implications for India', *Observer Research Foundation Occasional Papers* (January 25, 2022)

Joshi, M., *Understanding the India China Border. The Enduring Threat of War in the High Himalayas* (London: Hurst, 2022)

Joyner, D.H., *Interpreting the Nuclear Non-Proliferation Treaty* (Oxford: Oxford University Press, 2011)

Kelley, R.E., *Verifying Nuclear Disarmament. Lessons Learned in South Africa, Iraq and Libya* (Stockholm: SIPRI, January 2023)

Khan, F., 'Prospects for Indian and Pakistani Arms Control and CBMs', Nonproliferation Policy Education Center (February 24, 2010)

Kieninger, S., *George Shultz and the Road to the INF Treaty. Process and Personal Diplomacy* (Hoover Daily Report, December 11, 2020)

Kimball, D.G., 'Turning Point in the Struggle Against the Bomb: The Nuclear Ban Treaty Ready to Go into Effect', *Just Security* (October 27, 2020)

Kine, P., '"Spiral into Crisis" The U.S.–China Military Hotline Is Dangerously Broken', *Politico* (September 1, 2021)

Kirsten, I. and Zarka, M., *Balancing the Three Pillars of the NPT: How Can Promoting Peaceful Uses Help?*, EU Non-Proliferation and Disarmament Consortium, Non-Proliferation and Disarmament Papers, No. 79 (May 2022)

Kissinger, H., 'Should We Try to Defend Against Russia's Missiles?', *The Washington Post* (September 3, 1984) p C8

Knopf, J.W., 'Nuclear Disarmament and Nonproliferation. Examining the Linkage Argument', (winter 2012/13) 37(3) *International Security* 92

Kobayashi, Y., 'Observations on Lack of Transparency in China's Nuclear Arms Expansion: Ahead of the NPT Review Conference', *SPF China Observer*, no. 39 (August 17, 2022)

Koplow, D.A. and Ifft, E.M., 'Legal and Political Myths of the Treaty on the Prohibition of Nuclear Weapons', *Bulletin of the Atomic Scientists* (May 13, 2021)

Kristensen, H. and Korda, M., *The Treaty on the Prohibition of Nuclear Weapons Enters into Force Today* (Federation of American Scientists, January 22, 2021)

Kristensen, H.M. and Korda, M., 'Chinese Nuclear Weapons, 2021', (2021) 77(6) *Bulletin of the Atomic Scientists* 318

Kroenig, M., *Exporting the Bomb: Technology Transfer and the Spread of Nuclear Weapons* (Ithaca, NY: Cornell University Press, 2010)

Laikola, L. and Rolander, N., 'Nordic NATO Candidates Seek No Opt-Outs on Nuclear Weapons', *Bloomberg* (November 1, 2022)

LibreTexts Chemistry, 10.3: Isotopes of Hydrogen, https://chem.libretexts.org/Bookshelves/Inorganic_Chemistry/Map%3A_Inorganic_Chemistry_(Housecroft)/10%3A_Hydrogen/10.03%3A_Isotopes_of_Hydrogen

Logan, D.C. and Saunders, P.C., *Discerning the Drivers of China's Nuclear Force Development: Models, Indicators, and Data* (Washington, DC: National Defense University Press, July 2023)

Loose, H.-W., '2005—Year of the Nuclear Non-Proliferation Treaty—But What Happened to Nuclear Disarmament?' (2007) 4 *New Zealand Yearbook of International Law* 135

Mallard, G., 'Crafting the Nuclear Regime Complex (1950–1975): Dynamics of Harmonization of Opaque Treaty Rules', (2014) 25(2) *European Journal of International Law* 445, 449

Matheson, M.J., 'The Opinions of the International Court of Justice on the Threat or Use of Nuclear Weapons', (1997) 91(3) *American Journal of International Law* 417

McNeill, J.H., 'U.S.–USSR Nuclear Arms Negotiations: The Process and the Lawyer', (1985) 79 *American Journal of International Law* 52

Meyer, P., '"Permanence with Accountability": An Elusive Goal of the NPT', (2020) 3(2) *Journal for Peace and Nuclear Disarmament* 215

Miles, C.A., *Provisional Measures before International Courts and Tribunals* (Cambridge: Cambridge University Press, 2017)

Miller, S.E., 'Nuclear Hotlines: Origins, Evolution, Applications', (2021) 4 *Journal for Peace and Nuclear Disarmament* 176

Mizokami, K., 'Tactical Nuclear Weapons at Sea', *US Naval Institute Proceedings* (August 2020), www.usni.org/magazines/proceedings/2020/august/tactical-nuclear-weapons-sea

Moon, W.M., 'Beyond Arms Control: Cooperative Nuclear Weapons Reductions—a New Paradigm to Roll Back Nuclear Weapons and Increase Security and Stability', (2020) 3(1) *Journal for Peace and Nuclear Disarmament* 92

Morris, L.J. and Marcrum, K., 'Another "Hotline" With China Isn't the Answer', *TheRANDblog* (July 27, 2022)

Müller, H. and Wunderlich, C., 'Nuclear Disarmament without the Nuclear-Weapon States: The Nuclear Weapon Ban Treaty', (2020) 149(2) *Daedalus* 171

Nuclear Regulatory Commission (US), *Backgrounder on Tritium, Radiation Protection Limits, and Drinking Water Standards* (October 18, 2022), www.nrc.gov/reading-rm/doc-collections/fact-sheets/tritium-radiation-fs.html

Nye, Jr., J.S., 'The Logic of Inequality', (summer 1985) 59 *Foreign Policy* 123

Payne, K.B. and Costlow, M.R., *Deterring China: A Victory Denial Strategy*, National Institute for Public Policy, Information Series, Issue No. 519 (April 4, 2022), https://nipp.org/wp-content/uploads/2022/04/519.pdf

Peters, R., *It Is Time to Make the Next Generation of America's ICBMs Road-Mobile*, The Heritage Foundation, Issue Brief (January 11, 2024), www.heritage.org/defense/report/it-time-make-the-next-generation-americas-icbms-road-mobile

Pompeo, M.R. and Billingslea, M.S., 'China's Nuclear Build-Up Should Worry the West', *Newsweek* (January 4, 2021)

Popp, R., Horovitz, L., and Wenger, A. (eds.) *Negotiating the Nuclear Non-Proliferation Treaty: Origins of the Nuclear Order* (Abingdon, UK: Routledge, 2017)

Pretorius, J., 'Nuclear Politics of Denial: South Africa and the Additional Protocol', (2013) 18 *International Negotiation* 379

Rademaker, S., *U.S. Compliance with Article VI of the NPT*, Carnegie Endowment for International Peace (February 3, 2005)

Ramaker, J., Mackby, J., Marshall, P., and Geil, R., *The Final Test: A History of the Comprehensive Nuclear-Test-Ban Treaty Negotiations* (Vienna: Provisional Technical Secretariat of the Preparatory Commission for the Comprehensive Nuclear-Test-Ban Treaty Organization, 2003)

Richard, T., 'An Assessment of Russia's Withdrawal from the Comprehensive Test Ban Treaty', *Articles of War* (February 16, 2024), https://lieber.westpoint.edu/assessment-russias-withdrawal-comprehensive-test-ban-treaty/

Rogoff, M.A., 'The Obligation to Negotiate in International Law: Rules and Realities', (1994) 16(1) *Michigan Journal International Law* 141

Roscini, M., 'The Cases against the Nuclear Weapons States', (12 May 2015) 19(10) *ASIL Insights*: www.asil.org/insights/volume/19/issue/10/cases-against-nuclear-weapons-states#_ednref10

Ryan, M., 'The Pros and Cons of Hosting B-52s on Our Shores', *The Sydney Morning Herald* (November 1, 2022)

Schelling, T.C., *The Strategy of Conflict* (Cambridge: Harvard University Press, 1960)

Schelling, T.C., and Halperin, M.H., *Strategy and Arms Control* (New York: The Twentieth Century Fund, 1961)

Schneider, M.B., 'Yes, the Russians are Testing Nuclear Weapons and It Is Very Important', *RealClearDefense* (August 8, 2019)

Slatter, C., 'Securing the Pacific in a Globalised World: New and Emerging Developments in International Law', (2020) 27 *Canterbury Law Review* 5

Søe, C., 'Book Review: Jeffrey Herf. War by Other Means: Soviet Power, West German Resistance, and the Battle of the Euromissiles', (1995) 100(2) *American Historical Review* 497

Sokolski, H., 'Nuclear Non-Proliferation ... If You Can Keep It' (January 6, 2023), *ABA Standing Committee on Law & National Security*

Sokolski, H. and Beck, A., *Arresting Nuclear Adventurism: China, Article VI, and the NPT* (Nonproliferation Policy Education Center: Occasional Paper 2201, June 2022)

Sokolski, H. and Grant, T.D., 'China: A Nuclear Nonproliferation Deadbeat', *The Hill* (August 30, 2022)

Stricker, A., 'International Atomic Energy Agency', in R. Goldberg (ed.) *A Better Blueprint for International Organizations. Advancing American Interests on the Global Stage* (Foundation for the Defense of Democracy Press, Washington, DC: June 2021) 26

Suseanu, I., 'The NPT and IAEA Safeguards', (December 2021) 62(4) *IAEA Bulletin*

Sykes, L.R., *Silencing the Bomb: One Scientist's Quest to Halt Nuclear Testing* (New York: Columbia University Press, 2017)

Trachtenberg, D.J., Dodge, M., and Payne, K.B., *The 'Action-Reaction' Arms Race Narrative vs. Historical Realities* (Fairfax, VA: National Institute Press, March 2021)

Trump, D.J., *The Art of the Deal* (New York: Random House, 1987)

Tully, R., 'Will Donald Trump's Arms Control Policy Save Taiwan?' *RealClear Defense* (July 2, 2024)

UNIDIR, *A Fissile Material Cut-off Treaty. Understanding the Critical Issues* (2010), https://unidir.org/sites/default/files/publication/pdfs//a-fissile-material-cut-off-treaty-understanding-the-critical-issues-139.pdf#page=48

Valle Machado da Silva, M., 'Brazil and the Refusal to the Additional Protocol: Is It Time to Review This Position?' (2021) 16(1) *Carta Internacional* 1–26; 386–7

Van Cleave, W.R., 'The Arms Control Record: Successes and Failures', in R.F. Staar (ed.) *Arms Control: Myth Versus Reality* (Redwood City, CA: Stanford University Press, 1984) 3

von Hippel, F., 'Projecting Plutonium Stocks to 2040', in H. Sokolski (ed.) (March 2021) 65

Wang, H., Graff, G., and Dale-Huang A., *China's Growing Risk Tolerance in Space. People's Liberation Army Perspectives and Escalation Dynamics* (Santa Monica, CA: RAND Corporation, 2024)

Willrich, M., 'Safeguarding Atoms for Peace', (1966) 60(1) *American Journal for International Law* 34, 36–7

Wohlstetter, A., Jones G., and Wohlstetter, R., *Towards a New Consensus on Nuclear Technology*, Vol. 1, 1979, PH 78-04 - 832-33, a summary report prepared for the US Arms Control and Disarmament Agency, Contract #AC7 NC-106 (Los Angeles, CA: PanHeuristics, July 6, 1979) 34–5

Wright, T., 'Negotiations for a Nuclear Weapons Convention: Distant Dream or Present Possibility?', (2009) 10(1) *Melbourne Journal of International Law* 217

Xinjun, Z., 'Intentional Ambiguity and the Rule of Interpretation in Auto-Interpretation—The Case of "Inalienable Right" in NPT Article IV', (2009) 52 *Japanese Yearbook of International Law* 35

Yusuf, A.A., 'Engaging with International Law' (2020) 69 *International & Comparative Law Quarterly* 505

Zarate, R., 'The Three Qualifications of Article IV's "Inalienable Right"', in H. Sokolski (ed.) *Reviewing the Nuclear Nonproliferation Treaty* (Carlisle, PA: Strategic Studies Institute, May 2010) 219

Zhao, M., 'Is a New Cold War Inevitable? Chinese Perspectives on US–China Strategic Competition', (2019) 12(3) *Chinese Journal of International Politics* 371

Index

S

T

U

V

W